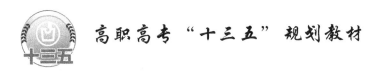
高职高专"十三五"规划教材

电工基础：电路分析
FOUNDATION OF ELECTRICAL ENGINEERING: ELECTRICAL CIRCUITS

（双语版）

主编　吕亚男　温贻芳

北　京
冶金工业出版社
2022

内 容 提 要

本书主要内容为电工技术的基本理论、电路定律以及分析方法，具体包括安全用电、电路元件、电路分析、电场与磁场、暂态电路分析、交流电和三相交流电等。本书对电工基础的内容进行了整理和精选，采用中英文对应，为一门电工基础类双语教材。

本书主要针对对象为高等职业教育机电类以及电子信息类专业等学生，也可用于机械类人员电工知识培训以及其他非电类专业的培训或供读者自学使用。

图书在版编目(CIP)数据

电工基础：电路分析/吕亚男，温贻芳主编.—北京：冶金工业出版社，2019.1（2022.1重印）
高职高专"十三五"规划教材
ISBN 978-7-5024-8008-0

Ⅰ.①电… Ⅱ.①吕… ②温… Ⅲ.①电路分析—高等职业教育—教材 Ⅳ.①TM133

中国版本图书馆 CIP 数据核字（2019）第 016360 号

电工基础：电路分析
FOUNDATION OF ELECTRICAL ENGINEERING：ELECTRICAL CIRCUITS

出版发行	冶金工业出版社		电　话	(010)64027926
地　　址	北京市东城区嵩祝院北巷39号		邮　编	100009
网　　址	www.mip1953.com		电子信箱	service@mip1953.com

责任编辑　卢　敏　美术编辑　吕欣童　版式设计　禹　蕊
责任校对　郑　娟　责任印制　李玉山

北京虎彩文化传播有限公司印刷
2019年1月第1版，2022年1月第2次印刷
787mm×1092mm　1/16；7.75印张；183千字；113页
定价 36.00 元

投稿电话　（010）64027932　投稿信箱　tougao@cnmip.com.cn
营销中心电话　（010）64044283
冶金工业出版社天猫旗舰店　yjgycbs.tmall.com
（本书如有印装质量问题，本社营销中心负责退换）

序(Preface)

《电工基础：电路分析》（Foundation of Electrical Engineering: Electrical Circuits）是高等职业教育机电类以及电子信息类等专业的专业基础课，目的是帮助高职学生理解及掌握电工技术的理论知识，包括定义概念、基本理论、电路定律以及分析方法等。

随着全球一体化的浪潮，我国高等职业教育也逐步趋向国际化和现代化，尤其中外合作办学项目对专业基础课的双语教学需求更为急迫。目前，电工基础课程的双语教学教材基本为原版教材或者双语教材，但主要针对本科及其以上的高等教育。对高等职业教育的专科学生而言，原版教材使用难度较大，适用的双语教材选择范围较小。因此，本书以高职教育对双语教材的需求为出发点，对电工基础内容进行了整理和精选，保证了必要的知识点和对应的英文内容。编制本教材时，本书采取了中文对基础概念、原理以及分析进行介绍，重点词汇用英文标注，重难点内容用英文进行总结梳理的方式；同时通过英文例题和练习题加强学生对电工基础知识的掌握和应用。

本书共7章，主要内容包括安全用电、电路元件、电路分析、电场与磁场、暂态电路分析、交流电和三相交流电等。内容详尽，表述清晰，可根据专业需求和课时需求进行选择性教学。

由于编者水平有限，书中难免存在一些缺点和不妥之处，欢迎读者批评指正，便于后期修订。

作者
2018年9月

目录(Contents)

1 安全用电(Electric Safety) ········· 1

  1.1 电力知识(Electric power knowledge) ········· 1
  1.2 安全用电(Electric safety) ········· 2
    1.2.1 安全电压(Safety voltage limit) ········· 2
    1.2.2 安全防护(Safety protection) ········· 2
  1.3 急救(First aid) ········· 2

2 电路元件(Electric Circuit Elements) ········· 4

  2.1 电路(Electric circuit) ········· 4
    2.1.1 电路基础(Circuit basic) ········· 4
    2.1.2 电路符号(Circuit symbol) ········· 4
  2.2 电流(Current) ········· 5
  2.3 电压(Voltage) ········· 6
  2.4 电功率和电能(Electric power and energy) ········· 6
  2.5 电阻与欧姆定律(Resistor and Ohm's Law) ········· 7
    2.5.1 电阻(Resistor) ········· 7
    2.5.2 电导(Conductance) ········· 7
    2.5.3 欧姆定律(Ohm's Law) ········· 8
  2.6 电压源与电流源(Voltage source and current source) ········· 9
    2.6.1 电压源(Voltage source) ········· 9
    2.6.2 电流源(Current source) ········· 10
    2.6.3 电源的等效变换(Source transformation) ········· 11
  2.7 支路(Branch) ········· 12
  2.8 节点(Node) ········· 12
  2.9 回路(Loop) ········· 12
  2.10 网孔(Mesh) ········· 12
  2.11 总结(Conclusion) ········· 13
  2.12 练习题(Exercises) ········· 13

3 电路分析(Circuits Analysis) ········· 16

  3.1 串联电路(Series circuit) ········· 16
  3.2 并联电路(Parallel circuit) ········· 19

目录（Contents）

- 3.3 混合电路（Hybrid circuit） 22
- 3.4 基尔霍夫定律（Kirchhoff's Law） 23
  - 3.4.1 基尔霍夫电流定律（Kirchhoff's Current Law） 23
  - 3.4.2 基尔霍夫电压定律（Kirchhoff's Voltage Law） 24
- 3.5 电路分析（Circuit analysis techniques） 26
  - 3.5.1 支路电流法（Branch current analysis） 27
  - 3.5.2 网孔分析法（Mesh current analysis） 28
  - 3.5.3 节点电压法（Nodal voltage analysis） 30
  - 3.5.4 叠加定理（Superposition theorem） 31
  - 3.5.5 戴维南定理（Thévenin's theorem） 33
  - 3.5.6 诺顿定理（Norton's theorem） 34
  - 3.5.7 三角形-星形等效变换（Delta-wye transformation） 36
- 3.6 练习题（Exercises） 37

4 电场与磁场（Electric Field and Magnetic Field） 43

- 4.1 电场（Electric field） 43
- 4.2 电容器和电容（Capacitor and capacitance） 43
  - 4.2.1 电容（Capacitance） 43
  - 4.2.2 电容器（Capacitor） 44
  - 4.2.3 电容器的连接（Connection of capacitor） 45
- 4.3 磁场与电磁场（Magnetic field and electromagnetic field） 47
- 4.4 电感元件和电感（Inductor and inductance） 47
  - 4.4.1 电感（Inductance） 47
  - 4.4.2 电感的连接（Connection of inductor） 49
- 4.5 练习题（Exercises） 50

5 暂态电路分析（Transient Circuit Analysis） 53

- 5.1 暂态过程与换路定律（Transient process and switching law） 53
  - 5.1.1 暂态过程（Transient process） 53
  - 5.1.2 换路定律（Switching law） 54
- 5.2 一阶 RC 电路的响应（First-order RC circuit） 55
  - 5.2.1 RC 电路的零输入响应（RC zero-input response） 55
  - 5.2.2 RC 电路的零状态响应（RC zero-state response） 57
  - 5.2.3 RC 电路的全响应（RC complete response） 58
- 5.3 一阶 RL 电路的响应（First-order RL circuit） 59
  - 5.3.1 RL 电路的零输入响应（RL zero-input response） 59
  - 5.3.2 RL 电路的零状态响应（RL zero-state response） 60
  - 5.3.3 RL 电路的全响应（RL complete response） 61
- 5.4 一阶电路的三要素法（Three-element method of the first order linear circuit） 62

5.5 练习题（Exercises） …………………………………………………………… 63

6 交流电（Alternating Current） ………………………………………………… 66

6.1 正弦交流电及其特征参数（AC and AC parameters） …………………… 66
6.1.1 正弦交流电（AC） ……………………………………………………… 66
6.1.2 特征参数（AC parameters） …………………………………………… 67
6.1.3 相量表示法（Phasor representation method） ……………………… 70
6.2 交流电路中的电阻（Resistor in AC circuit） …………………………… 72
6.2.1 电压和电流的关系（Relationship between voltage and current） …… 72
6.2.2 功率（Power） ………………………………………………………… 72
6.3 交流电路中的电容器（Capacitor in AC circuit） ……………………… 73
6.3.1 电压和电流的关系（Relationship between voltage and current） …… 73
6.3.2 功率（Power） ………………………………………………………… 74
6.4 交流电路中的电感（Inductor in AC circuit） …………………………… 76
6.4.1 电压和电流的关系（Relationship between voltage and current） …… 76
6.4.2 功率（Power） ………………………………………………………… 77
6.5 电阻、电容器和电感组成的电路（Resistor, capacitor and inductor in AC circuit） ……………………………………………………………… 78
6.5.1 阻抗（Impedance） …………………………………………………… 78
6.5.2 阻抗的串联和并联电路（Series and parallel circuit of impedance） … 81
6.6 谐振电路（Resonance circuit） …………………………………………… 83
6.6.1 串联谐振（Series resonance） ………………………………………… 83
6.6.2 并联谐振（Parallel resonance） ……………………………………… 86
6.7 功率因数的提高（Improvement of power factor） ……………………… 87
6.7.1 功率因数（Power factor） …………………………………………… 87
6.7.2 功率因数的提高（Improvement of power factor） ………………… 88
6.8 练习题（Exercises） ……………………………………………………… 89

7 三相交流电（Three-phase Sinusoidal AC Circuit） ………………………… 94

7.1 对称三相交流电源（Balanced three-phase voltage source） …………… 94
7.1.1 对称三相交流电压（Balanced three-phase voltage） ……………… 94
7.1.2 三相交流电源星形连接（Three-phase source wye connection） …… 96
7.1.3 三相交流电源三角形连接（Three-phase source delta connection） … 97
7.2 三相负载（Three-phase load） …………………………………………… 97
7.2.1 星形连接（Wye-wye circuit） ………………………………………… 97
7.2.2 三角形连接（Delta-delta circuit） …………………………………… 99
7.3 三相功率（Three-phase power） ………………………………………… 100
7.4 练习题（Exercises） ……………………………………………………… 101

8 附录（Appendix） ········ 103

8.1 Appendix One: Expanded material ········ 103
8.1.1 International system ········ 103
8.1.2 Electrical system ········ 103
8.1.3 Resistor ········ 104
8.1.4 Series and parallel connected circuits ········ 104
8.1.5 Alternating system ········ 105
8.2 Appendix Two: Basic electric quantities and units ········ 106
8.3 Appendix Three: Multiples/sub-multiples abbreviations ········ 107
8.4 Appendix Four: Mathematic representation ········ 108
8.5 Appendix Five: Greek letters ········ 109
8.6 Appendix Six: Vocabulary ········ 110

参考文献（References） ········ 113

1 安全用电（Electric Safety）

目标（Objectives）

In this chapter, you should
- be familiar with the electric power knowledge.
- recognize the electric safety.
- have an understanding of the first aid knowledge.

1.1 电力知识（Electric power knowledge）

电力系统包括发电厂发电、送电线路输电、变电线路变电、供配电所配电和用电设备用电等组成的电能生产与消耗系统。

发电的主要方式包括火力发电、水力发电、风力发电、核电和太阳能发电等。火力发电主要消耗的煤和石油等为不可再生资源且污染严重。水力发电、风力发电和太阳能发电为环保可再生资源发电。核电主要消耗核燃料，技术要求较高，所需安全等级高。

电力传输时，为了减少传输损耗，一般采用高压输电，我国常用输电电压为 $35\sim500kV$。

经过配电设施配电后，常用的生活生产电压为380V和220V。

Note：

(1) 发电（power generation）；
　　输电（power transmission）；
　　变电（power transformation）；
　　配电（power distribution）；
　　用电（power utilization）。
(2) 火力发电（thermal power generation）；
　　水力发电（hydroelectric generation）；
　　风力发电（wind power generation）；
　　核电（nuclear power generation）；
　　太阳能发电（solar power generation）。
(3) 不可再生资源（non-renewable resources）；
　　可再生资源（renewable resources）。
(4) 高压输电（high voltage transmission）。

1.2 安全用电（Electric safety）

1.2.1 安全电压（Safety voltage limit）

安全电压：对不佩戴任何防护设备的人体组织无法造成伤害的电压值称为安全电压。
安全电压标准：国际电工委员会：50V；
　　　　　　　中国：12V、24V、36V。
Note：
（1）安全电压标准（safety voltage standard）。
（2）国际电工委员会（International Electrical Commit）。

1.2.2 安全防护（Safety protection）

1.2.2.1 防护原则（Protection principle）

防护原则核心内容为安全工作、熟知并遵守安全守则、根据指示佩戴合格防护装备、注意安全标志、进行安全隐患排除。

1.2.2.2 防护装备（Protective device）

根据具体工种和工作环境，需要穿戴防护装备，如防护衣、安全帽、护目镜、防护面罩、防护手套等。

1.2.2.3 安全标志（Safety mark）

安全标志应明确清晰，通过图形和颜色的组合表示禁止、要求、预防、警告或急救。
禁止标志（prohibition）：严禁进行某种行为，如禁止饮食和禁止进入等。
要求标志（requirement）：规定某种行为，如戴安全帽和戴护目镜等。
警告标志（warning）：警示危险，如高空坠物警告和高温警告等。
急救标志（first-aid）：急救设备指示。

1.3 急救（First aid）

急救（first aid）：对伤者进行第一时间现场抢救的行为称为急救。
触电伤害（electric damage）：人体接触带电设备而使电流通过体表或者人体时造成的人体伤害（痉挛、窒息、心脏骤停甚至死亡）。
触电预防（precaution）：
（1）保证良好的绝缘条件（insulation condition）；
（2）对金属材料相关设备进行接地、接零处理（grounding processing）；
（3）带电设备要保持一定间距，必要时进行电气隔离（device electric isolation）。
对于触电伤害，应尽可能迅速展开现场急救。

首先，切断触电者身上的电流。

低电压设备（220V/380V）触电，可选用断开开关进行断电；若无法及时断电，需要立即通过不导电材料将线路或者设备与触电者分离。高压设备触电，需要第一时间进行呼救，只允许专业电气人员对电路进行断路操作，且受伤人员必须由专业人员进行救护，其余人员保持5m的安全距离。

其次，对触电伤害严重者，即呼吸心跳停止者，应马上进行人工呼吸和心脏按压急救处理；同时拨打紧急医疗救助电话，在急救的同时等待专业救护人员。

紧急医疗救助电话（emergency call）：120。

拨打120时，需要提供以下信息（information）：

何处？

何种事故？

受伤人数？

受伤程度？

2 电路元件（Electric Circuit Elements）

目标（Objectives）

In the chapter of electric circuit elements, you should
- have an understanding of the electric circuit current and voltage.
- be familiar with electric power and energy.
- have an understanding of the resistor and Ohm's law.
- be familiar with voltage source and current source.
- be capable of differentiating the branch, node, loop and mesh.

2.1 电路（Electric circuit）

2.1.1 电路基础（Circuit basic）

电能必须通过配电电路提供给用户。由电源（power source）、用电器（electric component）和导线（conductive line）等元器件连接组合而成作为电流的通路称为电路（electric circuit）。

电路通过开关的闭合与打开连接或断开电路，进而接通或断开用电器。电路具有三种状态：有载、开路和短路。

有载（loading status）：开关闭合（switch on），电源接通，与负载相互接通。

开路（open circuit）：开关断开（switch off），电源无法与负载接通，呈空载状态，电路电流为零。

短路（short circuit）：当电源两端无负载直接连接在一起时，称为电路短路。短路电流较大，会对电源设备造成损害。

Note：

（1）An electric circuit is a path in which electrons from a voltage or current source flow.

（2）An electric circuit is consist of power source, electric components and conductive lines.

（3）The connection and disconnection of an electric circuit are controlled by the switches.

2.1.2 电路符号（Circuit symbol）

电路图中用电路符号表示各电路设备（circuit elements）。电路符号仅表示电气设备的特性(characteristic),对电气设备的结构(structure)不做任何表示。表2-1为常用的电路符号。

表2-1 电路符号
Table 2-1 Electric circuit symbol

Term		Symbols
导线	conductive wire	——————
导线交叉	conductive crossover	—┼—
电源	power source	—┤├—
电阻	resistor	—▭—
电容	capacitor	—┤├—
电感	inductor	—⌒⌒⌒—
开关	switch	—/—

Note:
(1) A circuit diagram or wiring diagram uses symbols to represent the circuit elements.
(2) Symbols represent only the characteristics of the electricalelements, not their structures.
(3) The switch represents a control apparatus. The wires represent the transmission system.

2.2 电流(Current)

定义:单位时间内通过导体横截面的电荷(charge)量称为电流。

$$i = \frac{dq}{dt} \tag{2-1}$$

符号:电流大小和方向不随时间变化,称为直流电流,用 I 表示。
电流大小和方向随时间变化,称为交流电流,用 i 表示。
单位:安培,A。

Note:
(1) The electric current is the flow of electric charges.
(2) The unit of current is ampere, A.
(3) The electrons require a complete circuit to move.
(4) It is the source which provides a driving influence (electromotive force, emf) to make the current flow.
(5) Direct current (DC) is the flow of electricity in a single direction.
(6) Alternating current (AC) is an electric current of which magnitude and direction vary with the time.

Example 2-1:

In a circuit, if a charge of 100C flowing through a point in 20s, find I.

Solution:

$$I = \frac{Q}{t} = \frac{100}{20} = 5\text{A}$$

2.3 电压（Voltage）

定义：电压是电路中两点间的电位差（potential difference）。

电位称为电势（potential），是一个相对于参考点的物理量，数值上等于电场力（electric field force）将单位正电荷从某一点移动到参考点所做的功（work）。

如果电路中 a、b 两点的电位分别为 V_a 和 V_b，则

$$U_{ab} = V_a - V_b \tag{2-2}$$

即，电压值 U_{ab} 数值上等于电场力推动单位正电荷从 a 点移动到 b 点所做的功。

方向：由高电位（high potential）指向低电位（low potential）的方向，即为电压降低方向。为了方便计算，常将电压和电流的方向取一致的参考方向，称为关联参考方向（associated reference direction）。

符号：U。

单位：伏特，V；
    毫伏，mV；
    微伏，μV；
    千伏，kV。

Note:

(1) Voltage makes electric charges move.

(2) The unit of voltage is volt, V.

2.4 电功率和电能（Electric power and energy）

定义：电路或电路元件在单位时间内转换的电能（electrical energy）简称功率（power）。

符号：P。

$$P = UI \tag{2-3}$$

单位：瓦特，W；
    千瓦，kW；
    毫瓦，mW。

电路中单位时间内产生的电能：

$$W = Pt = UIt \tag{2-4}$$

单位：焦耳，J；
    常用 kW·h，千瓦·时，即度。

Note:

(1) Electric power is defined as the power dissipated by an electric circuit.

(2) Electric power is a measurement of the rate at which energy is used over a period of time.

(3) The unit for power is the watt.

The unit for energy is the joule.

The unit for time is the second.

(4) For a direct current circuit, electric power is equal to the prodnct of the elethic cunrent multipled by teh voltage.

Example 2-2:

Find the voltage of the load when a circuit delivers $P = 20W$ and $I = 10A$.

Solution:

$$U = \frac{P}{I} = \frac{20}{10} = 2V$$

2.5 电阻与欧姆定律（Resistor and Ohm's Law）

2.5.1 电阻（Resistor）

定义：对电流呈现阻碍作用的耗能元件的电路模型称为电阻。

符号：R。

单位：欧姆，Ω；

千欧，$k\Omega$；

兆欧，$M\Omega$。

分类：

线性电阻：电流随电压成比例变化的电阻。线性电阻可分为固定电阻，可变电阻和半可变电阻。

非线性电阻：电压和电流不成正比的电阻，如二极管。

电阻的图形符号如图 2-1 所示。

图 2-1 线性电阻（a）与非线性电阻（b）

Fig. 2-1 Symbol of linear resistor (a) and nonlinear resistor (b)

Note:

(1) A resistor implements electrical resistance.

(2) The unit of resistance is ohm, Ω.

(3) Variable resistors can adjust circuit elements or be used as sensing devices.

2.5.2 电导（Conductance）

定义：电阻的倒数称为电导。

符号：G

$$G = \frac{1}{R} \tag{2-5}$$

单位：西门子，S。

Note：

(1) Conductance is the reciprocal of resistance.

(2) The unit of conductance is siemens, S.

2.5.3 欧姆定律（Ohm's Law）

定义：施加于电阻元件上的电压与通过的电流成正比。当电路两端电压为1V，通过电流为1A时，电阻为1Ω。如图2-2所示。

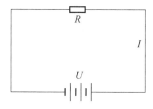

图 2-2 电路图

Fig. 2-2 Electric circuit diagram

$$R = \frac{U}{I} \tag{2-6}$$

where R——电阻（resistance），Ω；

U——电压（voltage），V；

I——电流（current），A。

Note：

(1) The resistance value of load is equal to the ratio of voltage to current provided that other physical factors remain unchanged.

(2) Gerog Simon Ohm (German, 1787—1854).

Example 2-3：

Find the resistance R in Fig. 2-3.

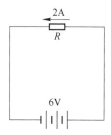

图 2-3 欧姆定律例题

Fig. 2-3 Example of Ohm's Law

Solution:

According to the Ohm's Law

$$R = \frac{U}{I} = \frac{6V}{2A} = 3\Omega$$

Example 2-4:

Find the power of the load and the heat dissipated in 10s. The current of 2A flows in the load with a resistance of 4Ω.

Solution:

$$P = I^2R = 2^2 \times 4 = 16W$$
$$W = Pt = 16 \times 10 = 160J$$

2.6 电压源与电流源（Voltage source and current source）

2.6.1 电压源（Voltage source）

电压源是一个理想的电路元件，其输出电压的大小和方向与外接电路无关，输出电压为定值或者给定的时间函数，流经电压源的电流由外接电路决定。

单位：伏特，V。

电压源图形符号如图 2-4 所示。

图 2-4 电压源符号

Fig. 2-4 Symbol of voltage source

Note:

(1) The voltage of the voltage source is constant (DC) or a time variable function (AC).

(2) The current flowing through the voltage source depends on the external circuit.

(3) The unit of voltage source is volt, V.

Example 2-5:

Find the current I in Fig. 2-5.

Solution:

According to the Ohm's Law,

$$I = \frac{U}{R} = \frac{6}{2} = 3A$$

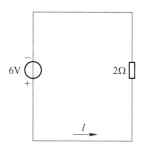

图 2-5 电压源例题

Fig. 2-5 Example of voltage source

2.6.2 电流源（Current source）

电流源的电流与其两端电压无关，电流值为定值或给定的时间函数，电流源两端的电压由其外接电路决定。

单位：安培，A。

电流源图形符号如图 2-6 所示。

图 2-6 电流源符号

Fig. 2-6 Symbol of current source

Note：

(1) The current flowing through the current source is constant.

(2) The unit of current source is ampere, A.

Example 2-6：

Find the voltage U in Fig. 2-7.

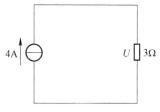

图 2-7 电流源例题

Fig. 2-7 Example of current source

Solution：

According to the Ohm's Law,

$$U = IR = 4 \times 3 = 12\text{V}$$

2.6.3 电源的等效变换 (Source transformation)

实际电压源和实际电流源之间可以进行等效变换。如图 2-8 所示。

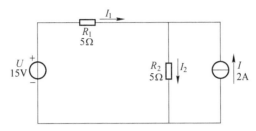

图 2.8 电源等效转换

Fig. 2-8　Example of source transformation

(1) 电压源等效转换为电流源模型时，R 保持不变：

$$I = \frac{U}{R}$$

转变后的等效电流源电路如图 2-9 所示。

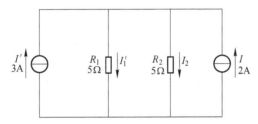

图 2-9　电压源转换为电流源

Fig. 2-9　Transformation of voltage source to current source

(2) 电流源转换为电压源模型时，R 保持不变：

$$U = IR$$

转变后的等效电压源电路如图 2-10 所示。

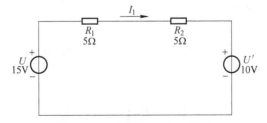

图 2-10　电流源转换为电压源

Fig. 2-10　Transformation of current source to voltage source

Note:

The voltage source and current source can be transformed between each other.

2.7 支路（Branch）

定义：电路中的每一条分支，叫作支路。

如图 2-11 所示，支路有 ab、$adcb$ 和 af。

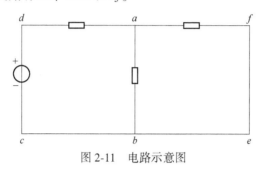

图 2-11 电路示意图

Fig. 2-11 Electrical schematic diagram

Note：

A branch is a single path in a network, composed of at least one simple element and the node at each end of that element.

2.8 节点（Node）

定义：电路中三条或三条以上的支路相连接的点，叫作节点。

如图 2-11 所示，a 和 b 为节点。

Note：

A node in a circuit is a point at which three or more circuit elements are joined together.

2.9 回路（Loop）

定义：电路中由一条或多条支路组成的闭合电路，叫作回路。

如图 2-11 所示，回路有 $abcd$、$abef$ 和 $dfec$。

Note：

A loop is a closed path starting at a node and proceeding through circuit elements, then returning to the starting node.

2.10 网孔（Mesh）

定义：电路中没有其他支路穿过的回路，叫作网孔。

如图 2-11 所示，网孔有 $abcd$ 和 $abef$。

Note：

Mesh is a loop where there is no branch across it.

2.11 总结(Conclusion)

(1) Current is a term to describe the flow of electric charge. Electric charge may be either positive or negative.

(2) The difference of the source and load in electric circuit is that the source supplies energy and the load accepts energy.

(3) The voltage is determined by the potential difference across a load.

(4) The electric power is the product of the current and the voltage.

(5) Based on the Ohm's law, the ratio of voltage to current is constant, and the ratio is equal to the resistance of the load.

2.12 练习题(Exercises)

2-1 Find the current I_1 and I_2 in Fig. 2-12.

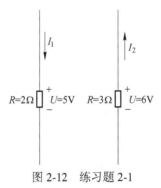

图 2-12 练习题 2-1

Fig. 2-12 Exercise 2-1

2-2 Find the voltage U_{ao}, U_{bo}, U_{co}, and U_{do} in Fig. 2-13.

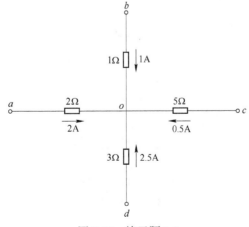

图 2-13 练习题 2-2

Fig. 2-13 Exercise 2-2

2-3 Find the voltage U_{ab}, U_{bc}, and U_{ca} in Fig. 2-14.

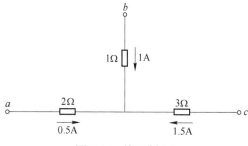

图 2-14 练习题 2-3
Fig. 2-14 Exercise 2-3

2-4 Find all the branches, nodes, loops and meshes in Fig. 2-15.

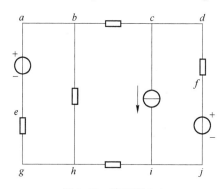

图 2-15 练习题 2-4
Fig. 2-15 Exercise 2-4

2-5 Find all the branches, nodes, loops and meshes in Fig. 2-16.

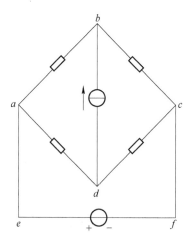

图 2-16 练习题 2-5
Fig. 2-16 Exercise 2-5

2-6 Find the voltage U_{ab} in Fig. 2-17.

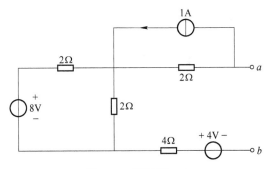

图 2-17　练习题 2-6
Fig. 2-17　Exercise 2-6

3 电路分析 (Circuits Analysis)

目标 (Objectives)

In this chapter, you should
- recognize the series, parallel and hybrid connected circuits.
- be capable of analyzing circuits with Kirchhoff's Law.
- have an understanding of the branch current analysis, mesh analysis, nodal analysis, superposition theorem, Thévenin's theorem, and Norton's theorem.

In Chapter 3, a variety of circuits analysis techniques are introduced to simplify and solve the circuit problems. Each technique owns the particular strength to solve particular circuit problem. The main techniques included are:
- Kirchhoff's Law
- Branch current analysis
- Mesh analysis
- Nodal analysis
- Superposition theorem
- Thévenin's theorem
- Norton's theorem

3.1 串联电路 (Series circuit)

定义：如果电路中有两个或者更多的电阻串在同一支路上，流过的电流相同，则称为电阻串联。

串联电路如图 3-1 所示。

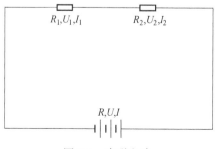

图 3-1 串联电路

Fig. 3-1 Series circuit

3.1 串联电路 (Series circuit)

串联电路中:

$$R = R_1 + R_2 + \cdots R_n$$
$$U = U_1 + U_2 + \cdots + U_n \qquad (3-1)$$
$$I = I_1 = I_2 = \cdots = I_n$$

Note:

(1) The current flowing through the components is same.

(2) Total resistance in series is the sum of the resistance of each component.

(3) Total voltage in series is the sum of the voltage of each component.

Example 3-1:

Find R in each circuit in Fig. 3-2.

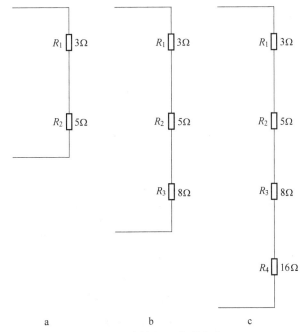

图 3-2 例题 3-1 串联电路

Fig. 3-2 Example 3-1 of series circuit

Solution:

Fig. 3-2a
$$R = R_1 + R_2 = 3 + 5 = 8\Omega$$

Fig. 3-2b
$$R = R_1 + R_2 + R_3 = 3 + 5 + 8 = 16\Omega$$

Fig. 3-2c
$$R = R_1 + R_2 + R_3 + R_4 = 3 + 5 + 8 + 16 = 32\Omega$$

Example 3-2:

Find current I in each circuit in Fig. 3-3.

Solution:

Fig. 3-3a:

$$I = \frac{U}{R} = \frac{220}{20} = 11\text{A}$$

Fig. 3-3b:

$$I = \frac{U}{R} = \frac{U}{R_1 + R_2} = \frac{220}{20 + 30} = 4.4\text{A}$$

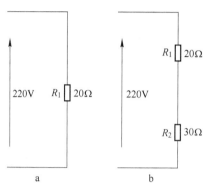

图 3-3 例题 3-2 串联电路

Fig. 3-3 Example 3-2 of series circuit

Example 3-3:

Find U_1, U_2, U_3 and I in the circuit shown in Fig. 3-4.

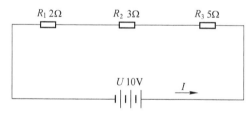

图 3-4 例题 3-3 串联电路

Fig. 3-4 Example 3-3 of series circuit

Solution:

$$R = R_1 + R_2 + R_3$$

According to the Ohm's law:

$$I = \frac{U}{R}$$

$$I = \frac{U}{R} = \frac{10}{2 + 3 + 5} = 1\text{A}$$

$$U_1 = IR_1 = 1 \times 2 = 2\text{V}$$
$$U_2 = IR_2 = 1 \times 3 = 3\text{V}$$
$$U_3 = IR_3 = 1 \times 5 = 5\text{V}$$

Example 3-4:

Find the R_1 in Fig. 3-5.

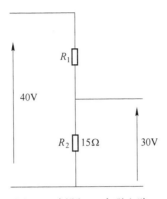

图 3-5 例题 3-4 串联电路

Fig. 3-5 Example 3-4 of series circuit

Solution:

In series circuit,

$$U = U_1 + U_2 = 40 = U_1 + 30$$

$$U_1 = 10V$$

$$U_1 = IR_1 = \frac{U_2}{R_2}R_1 = \frac{30}{15}R_1 = 10V$$

So

$$R_1 = 5\Omega$$

3.2 并联电路（Parallel circuit）

定义：如果电路中有两个或更多的电阻连接在两个公共结点之间，电阻上电压相同，则称为电阻并联。

并联电路如图 3-6 所示。

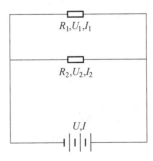

图 3-6 并联电路

Fig. 3-6 Parallel circuit

并联电路中，

$$\frac{1}{R} = \frac{1}{R_1} + \frac{1}{R_2} + \cdots + \frac{1}{R_n}$$

$$U = U_1 = U_2 = \cdots = U_n \tag{3-2}$$
$$I = I_1 + I_2 + \cdots + I_n$$

当电路中有两个电阻并联时,

$$I_1 = I \frac{R_2}{R_1 + R_2}$$

$$I_2 = I \frac{R_1}{R_1 + R_2}$$

Note:

(1) The voltage across each component is same.

(2) The reciprocal of total resistance in parallel is the sum of the reciprocal of each resistance.

(3) Total current in parallel is the sum of the current flowing through each branch.

Example 3-5:

Find R in each circuit in Fig. 3-7, $R_1 = 1\Omega$, and $R_2 = R_3 = R_4 = 2\Omega$.

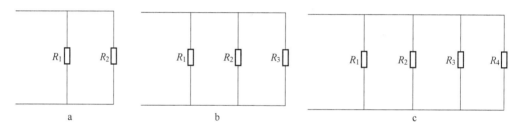

图 3-7 例题 3-5 并联电路

Fig. 3-7 Example 3-5 of parallel circuit

Solution:

Fig. 3-7a:

$$\frac{1}{R} = \frac{1}{R_1} + \frac{1}{R_2} = 1 + \frac{1}{2} = \frac{3}{2}$$

$$R = \frac{2}{3}\Omega$$

Fig. 3-7b:

$$\frac{1}{R} = \frac{1}{R_1} + \frac{1}{R_2} + \frac{1}{R_3} = 1 + \frac{1}{2} + \frac{1}{2} = 2$$

$$R = 0.5\Omega$$

Fig. 3-7c:

$$\frac{1}{R} = \frac{1}{R_1} + \frac{1}{R_2} + \frac{1}{R_3} + \frac{1}{R_4} = 1 + \frac{1}{2} + \frac{1}{2} + \frac{1}{2} = \frac{5}{2}$$

$$R = 0.4\Omega$$

Example 3-6:

Find U_1, U_2, U_3, I_1, I_2, I_3 and I in the circuit shown in Fig. 3-8.

3.2 并联电路 (Parallel circuit)

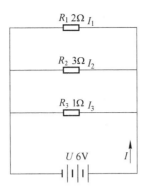

图 3-8 例题 3-6 并联电路

Fig. 3-8 Example 3-6 of parallel circuit

Solution:

$$U_1 = U_2 = U_3 = U = 6V$$

According to the Ohm's law,

$$I_1 = \frac{U_1}{R_1} = \frac{6}{2} = 3A$$

$$I_2 = \frac{U_2}{R_2} = \frac{6}{3} = 2A$$

$$I_3 = \frac{U_3}{R_3} = \frac{6}{1} = 6A$$

$$I = I_1 + I_2 + I_3 = 11A$$

Example 3-7:

In Fig. 3-9, find I_2 when R_1 is 1Ω, 2Ω and 3Ω, respectively.

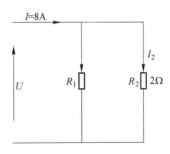

图 3-9 例题 3-7 并联电路

Fig. 3-9 Example 3-7 of parallel circuit

Solution:

When $R_1 = 1\Omega$,

$$I_2 = I \frac{R_1}{R_1 + R_2} = 8 \times \frac{1}{2+1} = 2.6A$$

When $R_1 = 2\Omega$,

$$I_2 = I\frac{R_1}{R_1 + R_2} = 8 \times \frac{2}{2+2} = 4\text{A}$$

When $R_1 = 3\Omega$,

$$I_2 = I\frac{R_1}{R_1 + R_2} = 8 \times \frac{3}{2+3} = 4.8\text{A}$$

3.3 混合电路（Hybrid circuit）

定义：同时包括串联电路和并联电路的电路称为混合电路。

对于复杂混合电路，需要对电路进行整理，计算电路中的等效电阻，确定总电压和总电流；根据电阻分压和分流关系，推算各支路电流或者电压。

混合电路如图3-10所示。

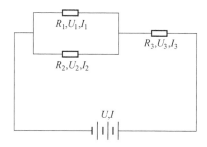

图 3-10　混合电路

Fig. 3-10　Hybrid integrated circuit

$$R = R_3 + R_{12}$$
$$\frac{1}{R_{12}} = \frac{1}{R_1} + \frac{1}{R_2}$$
$$R = R_3 + \frac{1}{\frac{1}{R_1} + \frac{1}{R_2}}$$
$$U = IR = U_1 + U_3 = U_2 + U_3$$
$$I = I_3 = I_1 + I_2$$

Note:

A hybrid integrated circuit is a miniaturized electronic circuit constructed of individual devices, such as semiconductor devices and passive components bonded to a substrate or printed circuit board.

Example 3-8:

In Fig. 3-10, $R_1 = 2\Omega$, $R_2 = 1\Omega$, $R_3 = \frac{1}{3}\Omega$, and $I_3 = 2\text{A}$, find U_1, U_2, U_3 and U.

Solution:

$$R = R_3 + \frac{1}{\frac{1}{R_1} + \frac{1}{R_2}} = \frac{1}{3} + \frac{1}{\frac{1}{2} + 1} = 1\Omega$$

$$U = IR = I_3 R = 2A \times 1\Omega = 2V$$

$$U_3 = I_3 R_3 = 2A \times \frac{1}{3}\Omega = 0.667V$$

$$U_1 = U_2 = U - U_3 = 2V - 0.667V = 1.333V$$

3.4 基尔霍夫定律（Kirchhoff's Law）

基尔霍夫定律（Kirchhoff's Law）是分析和计算电路的基本定律，包括基尔霍夫电压定律（KVL）和基尔霍夫电流定律（KCL）。

Gustav Kirchhoff found two general conditions named as Kirchhoff's Current Law and Kirchhoff's Voltage Law.

3.4.1 基尔霍夫电流定律（Kirchhoff's Current Law）

定义：任一瞬间，对电路中的任一节点，流入该节点的电流之和等于由该节点流出的电流之和。

基尔霍夫电流定律如图 3-11 所示。

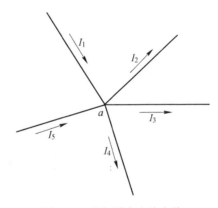

图 3-11 基尔霍夫电流定律

Fig. 3-11 Kirchhoff's Current Law

对节点 a，有

$$I_1 - I_2 - I_3 - I_4 + I_5 = 0$$

即 $\sum I = 0$，任一瞬时节点 a 电流的代数和恒等于零。

注意：

(1) 一般将流入节点的电流参考方向设为正向，流出节点的电流参考方向设为反向。

(2) KCL 可推广应用于包围部分电路的任一假设闭合面。

Note：

(1) KCL states the current balance of a node.

(2) The algebraic sum of the currents entering any node is zero.

(3) The sum of the currents entering a node is equal to the sum of the currents leaving a node.

(4) KCL is based on the principle of conservation of charge.

Example 3-9:

According to the KCL, determine the relationship between the currents I_1, I_2, I_3, I_4 and I_5 (Fig. 3-12).

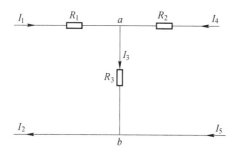

图 3-12 基尔霍夫电流定律例题

Fig. 3-12 KCL example

Solution:

For junction a,

$$I_1 + I_4 - I_3 = 0$$

For junction b,

$$I_3 + I_5 - I_2 = 0$$

Hence

$$I_1 + I_4 = I_2 - I_5$$

Or

$$I_1 - I_2 + I_4 + I_5 = 0$$

Example 3-10:

Find I_5 in Fig. 3-11, when $I_1 = 1\text{A}$, $I_2 = -2\text{A}$, $I_3 = 3\text{A}$, $I_4 = 1.5\text{A}$.

Solution:

According to the KCL,

$$I_1 - I_2 - I_3 - I_4 + I_5 = 0$$
$$1 - (-2) - 3 - 1.5 - I_5 = 0$$

Hence,

$$I_5 = 1.5\text{A}$$

3.4.2 基尔霍夫电压定律 (Kirchhoff's Voltage Law)

定义:任一瞬间,沿任一回路循环方向,电压降之和等于电压升之和,即回路中各段电压的代数和恒等于零。

对图 3-13 所示回路,有

$$U_1 + U_3 - U_{s1} = 0$$

3.4 基尔霍夫定律（Kirchhoff's Law）

$$U_2 + U_3 - U_{s2} = 0$$
$$U_1 - U_2 + U_{s2} - U_{s1} = 0$$

即
$$\sum U = 0 \tag{3-3}$$

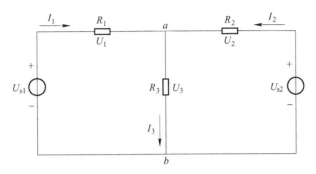

图 3-13　基尔霍夫电压定律

Fig. 3-13　Kirchhoff's Voltage Law

回路中，电压升降与绕行方向一致则为正，与绕行方向相反则为负。一般电流的参考方向取关联参考方向。

KVL 适用于闭合电路和开口电路。

Note：

(1) KVL states the voltage balance for a loop.

(2) The algebraic sum of the voltages around any closed path is zero.

(3) Travelling around a loop, the sum of the voltages rising is equal to the sum of the drooped voltages.

(4) KVL is based on the principle conservation of energy.

Example 3-11：

Find U_3 in Fig. 3-14.

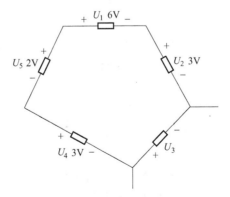

图 3.14　例题 3-11 基尔霍夫电压定律

Fig. 3-14　Example 3-11 of KVL

Solution：

According to the KVL,

$$U_1 + U_2 - U_3 - U_4 - U_5 = 0$$
$$6 + 3 - U_3 - 3 - 2 = 0$$

Hence,
$$U_3 = 4V$$

Example 3-12:

Find R_1, R_3 and U in Fig. 3-15, when the power dissipated in R_3 is 20 W.

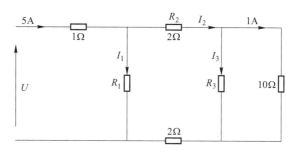

图 3-15 例题 3-12 基尔霍夫电压定律

Fig. 3-15 Example 3-12 of KVL

Solution:
$$P = U_3 I_3 = 10 \times 1 \times I_3 = 20W$$
$$I_3 = 2A$$

Also,
$$R_3 I_3^2 = 20W$$

Hence,
$$R_3 = 5\Omega$$

According to KCL,
$$I_2 = I_3 + 1 = 3A$$
$$I_1 = 5 - I_2 = 2A$$

According to KVL,
$$I_1 R_1 - 2 I_2 - I_3 R_3 - 2 I_2 = U_1 - 6 - 10 - 6 = 0$$
$$U_1 = 22V$$

Hence,
$$R_1 = \frac{U_1}{I_1} = \frac{22}{2} = 11\Omega$$

And
$$U + 5 \times 1 + U_1 = 0$$
$$U = -27V$$

3.5 电路分析 (Circuit analysis techniques)

基于基尔霍夫电流定律和基尔霍夫电压定律，开发了多种电路分析方法、如支路电流法、网孔分析法,节点电压法以及叠加定理法等,对电路进行分析计算。

3.5.1 支路电流法 (Branch current analysis)

方法:以支路电流为求解对象,应用基尔霍夫定律分别对节点和回路列出所需要的节点和回路方程,计算支路电流、电压和功率。

支路电流法如图3-16所示。

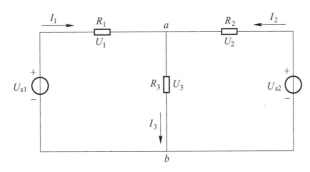

图 3-16 支路电流法

Fig. 3-16 Branch current analysis

步骤:

(1) 确定支路电流的参考方向。

(2) 设支路数量为 a,节点数量为 n。

(3) 根据 KCL,列节点的电流方程,n 个节点,可列 $n-1$ 个电流方程。

(4) 根据 KVL,列回路电压方程,数量为 $[a-(n-1)]$。

(5) 求解方程,得各支路电流。

如图 3-16 所示,支路数量为 3,节点数量为 2,则可列 1 个电流方程:

$$I_1 + I_2 = I_3$$

可列 2 个回路电压方程:

$$U_{s1} = I_1 R_1 + I_3 R_3$$
$$U_{s2} = I_2 R_2 + I_3 R_3$$

求解得 I_1、I_2、I_3。

Note:

(1) If there are N nodes and M meshes in a circuit, we can write $N-1$ KCL equations and $[M-(n-1)]$ KVL equations.

(2) Solving the equations to find the values of each branch current.

Example 3-13:

Find the current of each branch in the circuit in Fig. 3-17.

Solution:

There are four branches and two nodes.

The directions of current are supposed as shown in the Fig. 3-18.

According to the branch current analysis, the independent KCL equations are:

$$I_1 + I_3 = I_2$$
$$I_3 + 2 = I_4$$

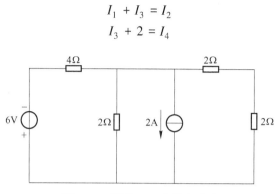

图 3-17 例题 3-17 支路电流法

Fig. 3-17 Example 3-17 of branch current analysis

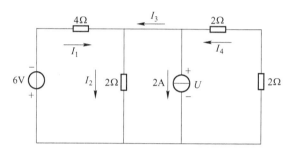

图 3-18 例题分析

Fig. 3-18 Example analysis

the independent KVL equations are：
$$6 + 4I_1 + 2I_2 = 0$$
$$U + 2I_4 + 2I_4 = 0$$
$$2I_2 - U = 0$$

After solving the equations, we have
$$I_1 = 0.625\text{A}, \ I_2 = -1.75\text{A}, \ I_3 = -1.125\text{A}, \ I_4 = 0.875\text{A}, \ U = -3.5\text{V}$$

3.5.2 网孔分析法（Mesh current analysis）

方法：以假想的网孔电流为未知量，应用 KVL，写出网孔方程，联立解出网孔电流，各支路电流则为有关网孔电流的代数和，这种分析方法称为网孔分析法。网孔法适用于平面电路。

步骤：

（1）标明各网孔电流及其参考方向。

（2）列出各网孔方程。电阻的正负取决于通过公共电阻的有关网孔电流的参考方向，一致时为正，否则为负。

（3）求解网孔方程，得到各网孔电流。

（4）选定各支路电流的参考方向，根据支路电流与网孔电流的线性组合关系，求得各电流。

(5) 利用元件的伏安关系,求得各支路电压。

如图 3-19 所示,假设网孔 1、2 的电流分别表示为 I_{m1} 和 I_{m2}。其参考方向如图 3-19 所示,选取绕行方向与网孔电流参考方向一致,则根据 KVL,可列出网孔方程:

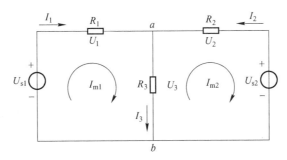

图 3-19 网孔分析法

Fig. 3-19 Mesh current analysis

网孔 1:
$$I_{m1} R_1 + (I_{m1} - I_{m2}) R_3 - U_{s1} = 0$$

网孔 2:
$$I_{m2} R_2 + (I_{m2} - I_{m1}) R_3 + U_{s2} = 0$$

整理得:
$$U_{s1} = I_{m1}(R_1 + R_3) - I_{m2}R_3$$
$$-U_{s2} = I_{m2}(R_2 + R_3) - I_{m1}R_3$$

Note:

(1) Mesh current analysis uses KVL equations to determine mesh currents in a circuit.

(2) For the circuit with N meshes, there are N independent KVL equations for N meshes.

(3) If a circuit contains current sources, the KVL equations of loops must avoid the current sources.

Example 3-14:

Find I_1 and I_2 in the circuit in the Fig. 3-20.

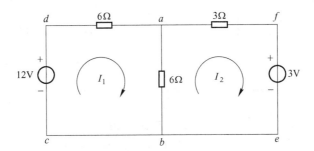

图 3-20 网孔分析法例题

Fig. 3-20 Example of mesh current analysis

Solution:

There are two meshes, mesh *abcd* and mesh *abef*.

According to the KVL equation:
$$6I_1 + 6(I_1 - I_2) - 12 = 0$$
$$3I_2 + 6(I_2 - I_1) + 3 = 0$$

Hence,
$$I_1 = 1.25A$$
$$I_2 = 0.5A$$

3.5.3 节点电压法（Nodal voltage analysis）

在具有 n 个节点的电路中，可以选其中一个节点作为基准，其余 ($n-1$) 个节点相对基准节点的电压称为节点电压。

方法：以节点电压为未知量，写出节点方程，从而解得节点电压，然后求出支路电流，这种分析方法叫节点电压法。

步骤：

（1）选定电路中任一节点为参考节点，用接地符号表示。

（2）标出各节点电压，其参考方向总是独立节点为"+"，参考节点为"-"。

（3）连接到本节点的电流源，当其电流指向节点时前面取正号；反之取负号。用观察法列出 ($n-1$) 个节点 KCL 电流方程。

（4）用 KVL 电压方程和欧姆定律求解节点方程，得到各节点电压。

（5）选定支路电流和支路电压的参考方向，计算各支路电流和支路电压。

如图 3-21 所示，节点 1、2、3 的节点电压分别为 U_{10}、U_{20} 和 U_{30}，则有：
$$U_1 = U_{10}$$
$$U_2 = U_{20}$$
$$U_3 = U_{30}$$
$$U_4 = U_{10} - U_{20}$$
$$U_5 = U_{30} - U_{20}$$

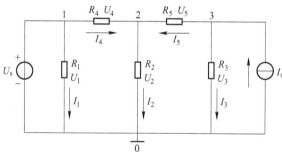

图 3-21 节点电压法

Fig. 3-21 Nodal voltage analysis

Note:

（1）Select one node as the ground-zero potential reference node and measure all other node voltages with respect to this node.

（2）Write down the KCL equations at the N-1 non-reference nodes.

(3) Use KVL equation and Ohm's Law to express the current by node voltages.

(4) Calculate the node voltage.

Example 3-15:

Find the U_{10}, U_{20} and U_{30} with nodal voltage analysis in Fig. 3-21. $U_s = 6V$, $I_s = 2A$, $R_1 = 6\Omega$, $R_2 = 1\Omega$, $R_3 = R_5 = 2\Omega$ and $R_4 = 3\Omega$.

Solution: According to KCL

$$\begin{cases} I_4 + I_5 = I_2 \\ I_3 + I_5 = 2 \end{cases}$$

$$I_2 = \frac{U_{20}}{R_2} = U_{20}$$

$$I_3 = \frac{U_{30}}{R_3} = \frac{U_{30}}{2}$$

$$I_4 = \frac{U_{10} - U_{20}}{R_4} = \frac{U_s - U_{20}}{R_4} = \frac{6 - U_{20}}{3}$$

$$I_5 = \frac{U_{30} - U_{20}}{R_5} = \frac{U_{30} - U_{20}}{2}$$

so $\quad U_{10} = 6V$, $U_{20} = 1.89V$, $U_{30} = 295V$

3.5.4 叠加定理 (Superposition theorem)

方法：在线性电路中，有几个独立电源共同作用时，每一个支路中产生的响应电流或电压，等于各个独立电源单独作用时在该支路中产生的响应电流或电压的代数和，这就是叠加定理。

如图 3-22 所示，则

$$U = U' + U''$$

$$U' = \frac{R_2 U_s}{R_1 + R_2}$$

$$U'' = I_2 R_2 = \frac{R_1 R_2 I_s}{R_1 + R_2}$$

$$U = \frac{R_2 U_s}{R_1 + R_2} + \frac{R_1 R_2 I_s}{R_1 + R_2} = \frac{R_2(U_s + R_1 I_s)}{R_1 + R_2}$$

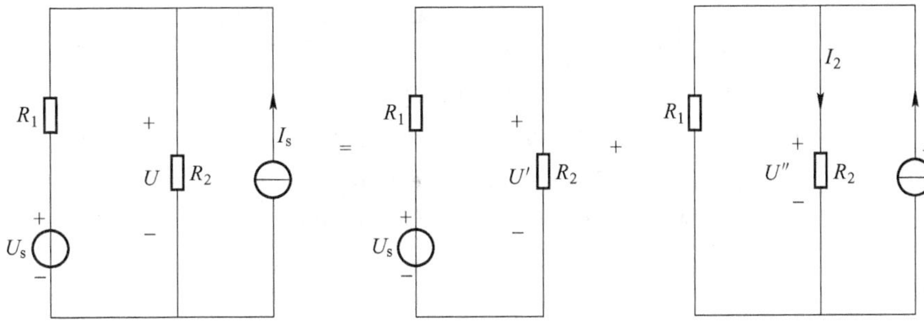

图 3-22 叠加定理

Fig. 3-22 Superposition theorem

叠加定理可用于计算线性电路的电压或电流，不能计算功率。在各个独立电源分别单独作用时，对那些暂不起作用的独立电源都应视为零值，即电压源用短路代替，电流源用开路代替，而其他元件的连接方式都不应有变动。

Note：

(1) In any linear resistive network, the voltage across or the current through any resistor or source may be calculated by adding algebraically all the individual voltages or currents caused by the separate independent sources acting alone.

(2) All other independent voltage sources are replaced by short circuits.

(3) All other independent current sources are replaced by open circuits.

(4) Superposition theorem works for the voltage and current calculation, not the power calculation.

Example 3-16：

Find I in Fig. 3-23.

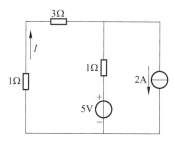

图 3-23　叠加定理例题

Fig. 3-23　Example of superposition theorem

Solution：

The source contribution circuits are shown in the Fig. 3-24.

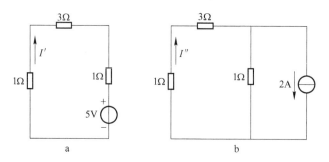

图 3-24　例题分析

a—电压源作用；b—电流源作用

Fig. 3-24　Example analysis

a—Voltage source active only；b—Current source active only

In Fig. 3-24a

$$I' = \frac{-5V}{1\Omega + 3\Omega + 1\Omega} = -1A$$

In Fig. 3-24b,

$$I'' = 2\text{A} \times \frac{1\Omega}{1\Omega + 3\Omega + 1\Omega} = 0.4\text{A}$$

Hence,

$$I = I' + I'' = -0.6\text{A}$$

3.5.5 戴维南定理 (Thévenin's theorem)

方法：任何一个有源的线性电阻单口网络，都可以等效为一个电压源和电阻串联的单口网络。

电压源的电压等于单口网络在负载开路时的电压，电阻是单口网络内全部独立电源为零值时所得单口网络的等效电阻，如图 3-25 所示。

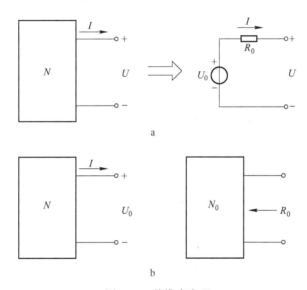

图 3-25 戴维南定理

Fig. 3-25 Thévenin's theorem

Note:

(1) Thévenin's theorem is named after M. Leon Thévenin (French, 1857—1926).

(2) It is possible to replace everything except the load resistor by an independent voltage source in series with a resistor.

(3) The response measured at the load resistor will be unchanged.

(4) The Thévenin source voltage is equal to the open-circuit voltage of two terminals network.

(5) The resistance is the equivalent resistance of two terminal network while all sources are zero.

Example 3-17:

Find U_0 in Fig. 3-26.

Solution:

The equivalent circuit is shown in the Fig. 3-27a, b and c.

In Fig. 3-27a,
$$U_{0c} = (3 \times 1) + (-4 \times 3) + (-1 \times 1) = -10V$$
In Fig. 3-27b,
$$R_{eq} = 3 + 1 + 1 = 5\Omega$$
In Fig. 3-27c, according to KVL,
$$U_0 = \frac{U_{0c}}{5+5} \times 5 = -5V$$

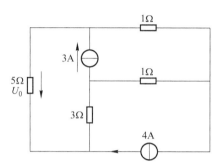

图 3-26 戴维南定理例题

Fig. 3-26 Example of Thévenin's theorem

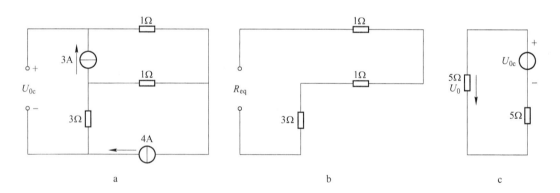

图 3-27 例题分析

a—等效电压；b—等效电源；c—等效电路

Fig. 3-27 Example analysis

a—equivalent voltage；b—equivalent resistance；c—equivalent circuit

3.5.6 诺顿定理 (Norton's theorem)

方法：任何一个有源二端线性网络，均可用一个理想电流源（I_s）和内阻（R_0）并联的电源来等效代替。

如图 3-28 所示，等效电源的电流为有源二端网络的短路电流，等效电源的电阻等于去掉网络中全部电源（理想电压源短路，理想电流源开路）后的等效电阻。

Note：

(1) Norton theorem is named after Edward Lawry Norton (American, 1898-1983).

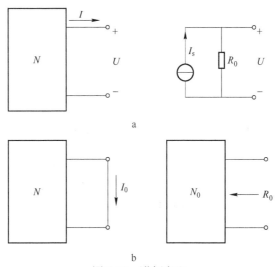

Fig. 3-28 Norton theorem

(2) It is an equivalent circuit that consists of a current source in parallel with a resistor.

Example 3-18:

Find I_0 in Fig. 3-29.

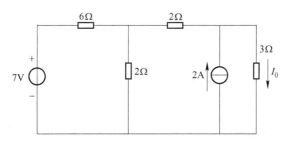

Fig. 3-29 Example of Norton theorem

Solution:

The equivalent circuit is shown in Fig. 3-30.

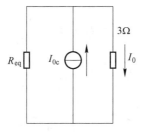

Fig. 3-30 Example analysis

$$I_{0c} = 2.5\text{A}$$

$$R_{eq} = 2 + \frac{1}{\frac{1}{6} + \frac{1}{2}} = 3.5\Omega$$

Hence,

$$I_0 = \frac{2.5 \times 3.5}{3.5 + 3} = 1.35\text{A}$$

3.5.7 三角形-星形等效变换 (Delta-wye transformation)

电阻连接常见的连接方式还有三角形连接 (Delta transformation, △) 和星形连接 (Wye transformation, Y)。如图 3-31 所示，星形连接中，三个电阻有共同端点，另一端端钮不同。而在三角形连接中，三个电阻端钮首尾相接。

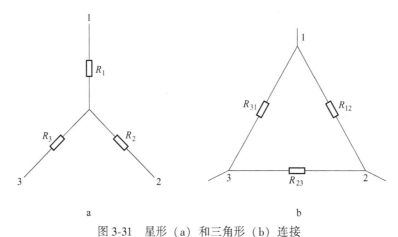

图 3-31 星形 (a) 和三角形 (b) 连接

Fig. 3-31 Wye connection (a) and delta connection (b)

若图中端钮 1、2、3 的电流对应相等且端钮间电压对应相等，则三角形连接和星形连接可以进行等效转化。需要注意的是，两者端点之间的电阻不可以直接采用原连接方式的阻值，需转换计算。

星形-三角形转换：

$$R_{12} = R_1 + R_2 + \frac{R_1 R_2}{R_3}$$

$$R_{23} = R_2 + R_3 + \frac{R_2 R_3}{R_1} \tag{3-4}$$

$$R_{31} = R_1 + R_3 + \frac{R_1 R_3}{R_2}$$

三角形-星形转换：

$$R_1 = \frac{R_{31} R_{12}}{R_{12} + R_{23} + R_{31}}$$

$$R_2 = \frac{R_{12} R_{23}}{R_{12} + R_{23} + R_{31}} \tag{3-5}$$

$$R_3 = \frac{R_{23} R_{31}}{R_{12} + R_{23} + R_{31}}$$

3.6 练习题（**Exercises**）

3-1 Find the equivalent resistance R_{eq} in Fig. 3-32.

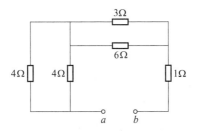

图 3-32 练习题 3-1

Fig. 3-32 Exercise 3-1

3-2 Find the equivalent resistance R_{eq} in Fig. 3-33.

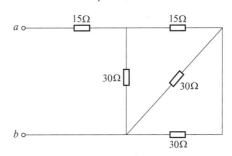

图 3-33 练习题 3-2

Fig. 3-33 Exercise 3-2

3-3 Find the voltage U_1 and U_2 in Fig. 3-34.

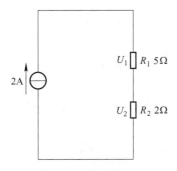

图 3-34 练习题 3-3

Fig. 3-34 Exercise 3-3

3-4 Find the current I in Fig. 3-35.

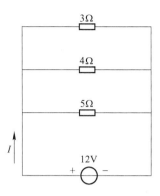

图 3-35　练习题 3-4

Fig. 3-35　Exercise 3-4

3-5　Find U_{AB} in the circuit shown in Fig. 3-36.

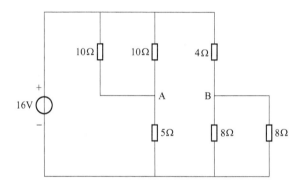

图 3-36　练习题 3-5

Fig. 3-36　Exercise 3-5

3-6　Find the current I and voltage U using branch current analysis in Fig. 3-37.

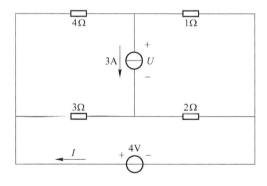

图 3-37　练习题 3-6

Fig. 3-37　Exercise 3-6

3-7　Find the current I_1, I_2 and I_3 using branch current analysis in Fig. 3-38.

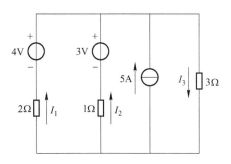

图 3-38 练习题 3-7
Fig. 3-38 Exercise 3-7

3-8 Find the current of each branch using mesh current analysis in Fig. 3-39.

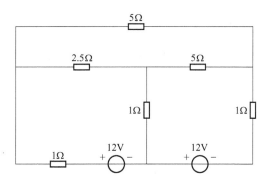

图 3-39 练习题 3-8
Fig. 3-39 Exercise 3-8

3-9 Find U_0 using mesh current analysis in Fig. 3-40.

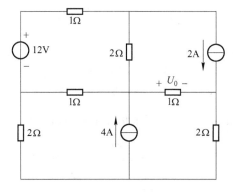

图 3-40 练习题 3-9
Fig. 3-40 Exercise 3-9

3-10 Find U_1 and U_2 using nodal voltage analysis in Fig. 3-41.

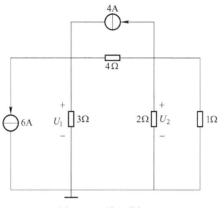

图 3-41 练习题 3-10
Fig. 3-41 Exercise 3-10

3-11 Using the Nodal Voltage method calculate the voltages V_1 and V_2 in Fig. 3-42.

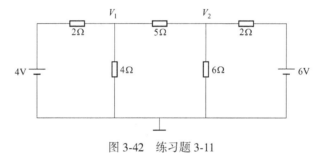

图 3-42 练习题 3-11
Fig. 3-42 Exercise 3-11

3-12 Find U using superposition principle in Fig. 3-43.

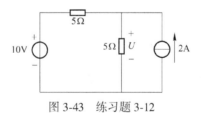

图 3-43 练习题 3-12
Fig. 3-43 Exercise 3-12

3-13 Find U using superposition principle in Fig. 3-44.

图 3-44 练习题 3-13
Fig. 3-44 Exercise 3-13

3-14 Find the Thévenin's equivalent in Fig. 3-45.

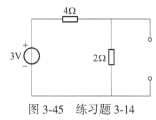

图 3-45 练习题 3-14
Fig. 3-45 Exercise 3-14

3-15 Find the current I with Thévenin's theorem in Fig. 3-46.

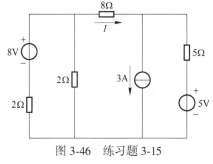

图 3-46 练习题 3-15
Fig. 3-46 Exercise 3-15

3-16 Find the Norton equivalent in Fig. 3-47.

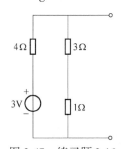

图 3-47 练习题 3-16
Fig. 3-47 Exercise 3-16

3-17 Find I using Norton equivalent in Fig. 3-48.

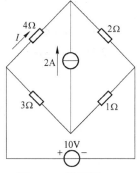

图 3-48 练习题 3-17
Fig. 3-48 Exercise 3-17

3-18 Find *I* in Fig. 3-49 using (a) Kirchhoff's laws and (b) Thévenin's theorem.

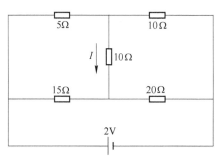

Fig. 3-49 Exercise 3-18

3-19 Find the voltage across the 4Ω resistor in Fig. 3-50 using (a) Nodal analysis, (b) the Superposition theorem, and (c) Thévenin's theorem.

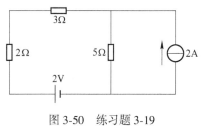

Fig. 3-50 Exercise 3-19

3-20 Find the equivalent resistance R_{AB} and R_{AN} in Fig. 3-51.

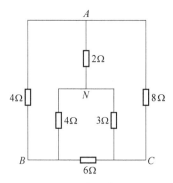

Fig. 3-51 Exercise 3-20

4 电场与磁场（Electric Field and Magnetic Field）

目标（Objectives）

In the chapter of electric field and magnetic field, you should
- be familiar with the electric field.
- have an understanding of capacitor and capacitance.
- be familiar with the magnetic field.
- have an understanding of inductor and inductance.
- be capable of applying series and parallel-connected circuits of capacitors and inductors.

4.1 电场（Electric field）

在带正负电的电极之间，存在一个对电荷有力作用的电场。

电场中，作用在电荷（electric charge）上的力（force）与电荷大小成正比，作用在单位电荷上的力的强弱称为电场强度（electric field intensity）。

电场强度用 E 表示：

$$E = \frac{F}{Q} \tag{4-1}$$

where　E——电场强度，牛顿/库仑，N/C；

F——力，牛顿，N；

Q——电荷，库仑，C。

Note：

(1) An electric field is a vector field surrounding an electric charge that exerts force on other charges, attracting or repelling them.

(2) Electric field intensity is equal to the ratio of force to electric charge.

4.2 电容器和电容（Capacitor and capacitance）

4.2.1 电容（Capacitance）

电容元件的储能本领可用电容 C 表示：

$$C = \frac{Q}{U} \tag{4-2}$$

where C——电容,法拉,F;
　　　Q——电荷量,库仑,C;
　　　U——电压,伏特,V。
　电容单位:法拉,F
　　　　　毫法,mF,$1mF = 10^{-3}F$
　　　　　微法,μF,$1\mu F = 10^{-6}F$
　　　　　纳法,nF,$1nF = 10^{-9}F$
　　　　　皮法,pF,$1pF = 10^{-12}F$

Note:

(1) Capacitance is the ability to store power energy.

(2) The unit of capacitance is farad, F.

Example 4-1:

A voltage of 100V is applied to a capacitor with a capacitance of 250μF. Find the electrical charge.

Solution:

According to equation (4-2)

$$Q = UC = 100 \times 250 \times 10^{-6} = 0.025C$$

4.2.2 电容器 (Capacitor)

电容器由在其中间带有绝缘材质的两个金属导电板和电介质组成。电容器可储存电荷。电容器的电流与电压的变化率成正比,具有隔直流通交流的作用。

电容器电气元件如图 4-1 所示。

图 4-1　电容符号

Fig. 4-1　Symbol of capacitor

常用电容器金属材料见表 4-1。

表 4-1　电容器常用金属材料

Table 4-1　Typical materials of capacitor

No. (序号)	1	2	3	4	5	6
Material (材料)	copper (铜)	aluminum (铝)	silver (银)	platinum (铂)	bronze (青铜)	gold (金)

常用绝缘材料见表 4-2。

表 4-2 常用绝缘材料
Table 4-2 Typical insulating materials

No. (序号)	Material (材料)	Relative permittivity (相对介电常数)
1	vacuum (真空)	1.0
2	air (空气)	1.0006
3	paper (dry) (纸)	2~2.5
4	rubber (橡胶)	2~3.5
5	insulating oil (绝缘油)	3~4
6	glass (玻璃)	5~10

电容元件上的电压、电流关系 (relation of current and voltage of the capacitor):

$$i_C = C\frac{du}{dt} \tag{4-3}$$

功率 (power):

$$p = ui = Cu\frac{du}{dt} \tag{4-4}$$

能量 (energy):

$$Wc(t) = \frac{1}{2}Cu^2(t) \tag{4-5}$$

Note:

(1) Capacitor can store power energy.

(2) Capacitor is consist of two metallic plate and insulating material.

(3) The current of capacitor is proportional to the change rate of voltage.

(4) Capacitors can be connected in series and in parallel.

4.2.3 电容器的连接 (Connection of capacitor)

在电路中,电容器可以串联 (series connection)、并联 (parallel connection) 和串并联 (hybrid connection) 混合接入电路。

4.2.3.1 电容器的串联 (Series connection of capacitors)

如图 4-2 所示,串联电路中,流经各电容器的电流相等,总电容小于单个电容;单个电容器均具有分电压作用,其值与电容成反比。

电容器串联电路的等效电容为:

$$\frac{1}{C} = \frac{1}{C_1} + \frac{1}{C_2} \tag{4-6}$$

Note:

(1) The current flowing through each capacitor is the same.

(2) The total capacitance is less than each capacitance.

(3) The reciprocal of total capacitance in series is the sum of the reciprocal of each capacitance.

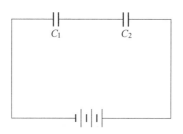

图 4-2 电容的串联

Fig. 4-2 Series circuit of capacitors

4.2.3.2 电容器的并联 (Parallel connection of capacitors)

如图 4-3 所示,并联电路中,各电容的电压相等;总电容为单个电容的和。

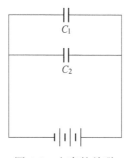

图 4-3 电容的并联

Fig. 4-3 Parallel circuit of capacitors

电容器并联电路的等效电容为:

$$C = C_1 + C_2 \tag{4-7}$$

Note:

(1) The voltage across each capacitor is the same.

(2) The total capacitance is the sum of each capacitance.

Example 4-2:

Three capacitors are with capacitances of 1F, 2F and 4F. Find the equivalent capacitance when they are connected in (a) series and (b) parallel.

Solution:

In series:

$$\frac{1}{C} = \frac{1}{C_1} + \frac{1}{C_2} + \frac{1}{C_3} = 1 + \frac{1}{2} + \frac{1}{4} = \frac{7}{4}\text{F}$$
$$C = 0.57\text{F}$$

In parallel:
$$C = C_1 + C_2 + C_3 = 1 + 2 + 4 = 7\text{F}$$

Example 4-3:

Find the equivalent capacitance C in Fig. 4-4.

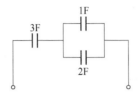

图 4-4 例 4-1

Fig. 4-4 Example 4-1

Solution:
$$\frac{1}{C} = \frac{1}{3} + \frac{1}{1+2} = \frac{2}{3}\text{F}$$
$$C = 1.5\text{F}$$

4.3 磁场与电磁场 (Magnetic field and electromagnetic field)

磁体中及其周围的空间有一个能量场，称为磁场。

电流流过的导体产生的磁场叫作电磁场。

磁场方向与电流方向有关，磁场强度随距导线距离增加而减弱。

Note:

(1) A magnetic field is a vector field that describes the magnetic influence of electrical currents and magnetized materials.

(2) An electromagnetic field is a physical field produced by electrically charged objects. The field can be viewed as the combination of an electric field and a magnetic field.

(3) The direction of magnetic field is related to the direction of current.

4.4 电感元件和电感 (Inductor and inductance)

4.4.1 电感 (Inductance)

由多个串联的导体线匝组成的元件叫作电感 (Inductor)。

电流通过线圈时产生电磁场，一个线圈的磁场由每个线匝磁场叠加而成：

$$L = \frac{N\phi}{i} \tag{4-8}$$

where　L——自感系数或电感，亨利，H；
　　　　ϕ——磁通量，韦伯，Wb；
　　　　N——线圈匝数，n；
　　　　i——通过电感的电流，安培，A。
电感符号如图 4-5 所示。

图 4-5　电感符号

Fig. 4-5　Symbol of inductor

电感元件电压与电流的关系：

线圈中通以电流 i，将会产生磁通 Φ，若磁通 Φ 与线圈的 N 匝都交链，则磁通链 $\psi = N\Phi$。若线圈中电流和端电压为关联参考方向，则：

$$u(t) = L \frac{\mathrm{d}i}{\mathrm{d}t} \tag{4-9}$$

电感的特性：
1）电感上电压与电流的变化率成正比，电感元件具有通直流的作用。
2）电感上的电流只能连续变化，不能跃变。
3）电感上的电流具有记忆过去电压的作用。
4）电感元件的功率与能量：

功率 (power)：

$$p = ui = Li \frac{\mathrm{d}i}{\mathrm{d}t} \tag{4-10}$$

能量 (energy)：

$$W_L(t) = \frac{1}{2} L i^2(t) \tag{4-11}$$

Note：

(1) The voltage of an inductor is proportional to the change of the current and opposes the change in current.

(2) The unit of inductance is henry, H (Joseph Henry, American physicist, 1797 - 1878).

(3) The stored energy of an inductor depends on the current.

Example 4-4：

Find the flux of the coil produced by the current of 50A, when the coil N of 200 is wound on a non-magnetic core, and the inductance is 10mH.

Solution：

According to the equation (4-8)：

$$L = \frac{N\phi}{i}$$

$$\phi = \frac{Li}{N} = \frac{10 \times 10^{-3} \times 50}{200} = 2.5 \times 10^{-3} \mathrm{Wb}$$

Example 4-5:

A coil is wound with 200 turns and the flux is 400μWb. Determine the inductance of the coil when the current is (a) 2A, and (b) 10A.

Solution:

According to the equation (4-8),

for $i = 2$A:

$$L = \frac{N\phi}{i} = \frac{200 \times 400 \times 10^{-6}}{2} = 0.04\text{H}$$

for $i = 10$A:

$$L = \frac{N\phi}{i} = \frac{200 \times 400 \times 10^{-6}}{10} = 0.008\text{H}$$

4.4.2 电感的连接（Connection of inductor）

在电路中，电感器可以串联（series connection）、并联（parallel connection）和串并联（hybrid connection）混合接入电路。

4.4.2.1 电感的串联（Series connection of inductors）

如图 4-6 所示，串联电路中，总电感为单个电感的和。

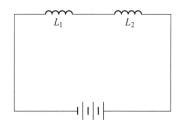

图 4-6 电感的串联

Fig. 4-6 Series circuit of inductors

电感器串联电路的等效电容为：

$$L = L_1 + L_2 \tag{4-12}$$

Note:

(1) The current flowing through each inductor is the same.

(2) The total inductance is the sum of each inductance.

4.4.2.2 电感的并联（Parallel connection of inductors）

如图 4-7 所示，并联电路中，流经各电容器的电流相等，总电容小于单个电容；单个电容器均具有分电压，其值与电容成反比。

并联的等效电感为：

$$\frac{1}{L} = \frac{1}{L_1} + \frac{1}{L_2} \tag{4-13}$$

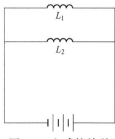

图 4-7 电感的并联

Fig. 4-7 Parallel circuit of inductors

Note:

(1) The voltage applied to each inductor is the same.

(2) The total inductance is less than each inductance.

(3) The equations of inductors apply to more than two inductors.

4.5 练习题 (Exercises)

4-1 Three capacitors are connected in (a) series and (b) parallel. The capacitances are 5F, 10F and 20F, respectively. Find the equivalent capacitance C.

4-2 Find the equivalent capacitance C in Fig. 4-8.

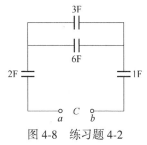

图 4-8 练习题 4-2

Fig. 4-8 Exercise 4-2

4-3 Find the equivalent capacitance C in Fig. 4-9.

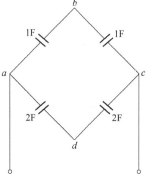

图 4-9 练习题 4-3

Fig. 4-9 Exercise 4-3

4-4 Find the equivalent capacitance C in Fig. 4-10.

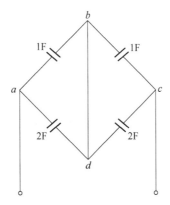

图 4-10 练习题 4-4
Fig. 4-10 Exercise 4-4

4-5 Two capacitors of 2mF and 5mF are connected in series with a power source of 220V. Determine (a) the equivalent capacitance, (b) the charge on each capacitor, and (c) the voltage across each capacitor.

4-6 A capacitance of 2.5F is required by the combination of two capacitors. The capacitance of capacitor (a) is 3F. What is the capacitance of another capacitor, and how to combine them?

4-7 Calculate the equivalent capacitance when three inductors are connected in (a) series and (b) parallel. The inductances are 5H, 10H and 20H, respectively.

4-8 An inductance of 2.5H is required by the combination of two inductors. The inductance of inductor (a) is 3H. What is the inductance of another inductor, and how to combine them?

4-9 Find the equivalent capacitance L in Fig. 4-11.

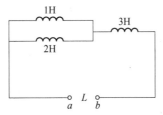

图 4-11 练习题 4-9
Fig. 4-11 Exercise 4-9

4-10 R is a pure resistor, L is an inductor, and C is a capacitor in Fig. 4-12. The light bulbs a, b and c are with same specifications. When the switch is turned on, find the orders that the light on of the three bulbs.

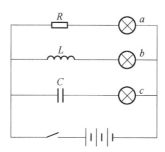

图 4-12　练习题 4-10
Fig. 4-12　Exercise 4-10

5 暂态电路分析（Transient Circuit Analysis）

目标（Objectives）

When you have studied this chapter, you should
- recognize the transient response of RC and RL circuits.
- have an understanding of the response of L and C to frequency variation.
- be capable of analyzing the response of an RLC series circuit to frequency variation.
- be capable of analyzing the first-order circuit.
- be familiar with the three-element method.

在只有电阻元件的电路中，接通或断开电源，电路会瞬间达到稳定状态，称为稳态电路。

当电路中有电容或电感元件时，电路并不会随着电源接通或断开而瞬时达到稳态，即出现暂态电路。

Note:

(1) If there are only resistors in the circuit, the responses are reaching steady state instantly when the power is on or off.

(2) If there are capacitors or inductors in the circuit, the steady status will not achieve instantly when the power is on or off.

5.1 暂态过程与换路定律（Transient process and switching law）

5.1.1 暂态过程（Transient process）

暂态过程是由储能元件（电感和电容）存储的能量不能跃变导致的。

定义：

含有电感或电容元件的电路中，在两个稳定状态中间，电流和电压从变化趋于稳定的过渡过程叫作暂态过程。

Note:

The transient status is an interim process from one steady state to another steady state, and the process is not instant.

5.1.2 换路定律 (Switching law)

定义：

引起电路工作状态发生变化的因素，叫作换路。换路包括开关闭合打开、短路、电压改变、电路参数改变等。换路过程是稳态和暂态的连接点。

如图 5-1 所示，假设换路瞬间为 $t = 0$，

则 $t = 0_-$ 称为换路前终了瞬间；

$t = 0_+$ 称为换路后的初始瞬间；

$t = \infty$，电路达到稳态。

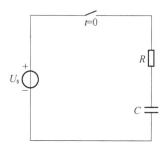

图 5-1 RC 电路

Fig. 5-1 RC circuit

因此，$t = 0_-$ 和 $t = \infty$ 时，电路为稳态。稳态电路中，电感元件视为短路，电容元件视为开路；

$t = 0_+$ 时，电容和电感为储能元件，能量不能跃变，此时

$$u_C(0_-) = u_C(0_+) \tag{5-1}$$

$$i_L(0_-) = i_L(0_+) \tag{5-2}$$

即为换路定律，用于确认电路暂态过程的初始值。

Note：

(1) All the factors leading to the variation of the operation status are called switching, such as power on or off, short circuit, and the change of the voltage, etc.

(2) Switching process is the joint of the steady and the transient state.

(3) The resistor does not store energy, so it does not follow the switching rule.

(4) If $t = 0$ represents the moment of switching,

$t = 0_-$ represents the ending point before switching;

$t = 0_+$ represents the initial point after switching;

$t = \infty$ represents the circuit is back to steady.

Example 5-1：

In Fig. 5-2, assuming the switch is turned on for a long timed, at $t = 0$, the switch is turned off. Find the i_C, i_L, and u_C.

Solution：

At $t = 0$, a short circuit is instead of the inductor, and an open circuit is instead of the capac-

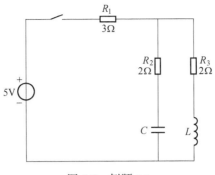

图 5-2 例题 5-1

Fig. 5-2　Example 5-1

itor, as shown in the Fig. 5-3.

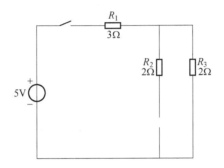

图 5-3 例题分析

Fig. 5-3　Example analysis

$$i_C = 0\text{A}$$
$$i_L = \frac{5\text{V}}{2\Omega + 3\Omega} = 1\text{A}$$
$$u_C = \frac{2\Omega}{2\Omega + 3\Omega} \times 5\text{V} = 2\text{V}$$

5.2　一阶 RC 电路的响应（First-order RC circuit）

5.2.1　RC 电路的零输入响应（RC zero-input response）

RC 零输入响应定义为，电路在无激励下输入信号为零，由电容释放储能的放电过程。

如图 5-4 所示，换路前，开关位置为 1，电源对电容充电；$t = 0$ 时，开关位置为 2，此时无电源提供能量，输入信号为零，电容将储有的能量释放，其电压初始值为 $u_C(0_+) = U_0$。

根据 KVL 得：

$$RC\frac{\mathrm{d}u_C}{\mathrm{d}t} + u_C = 0 \tag{5-3}$$

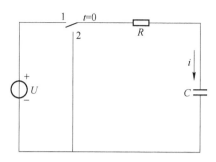

图 5-4 RC 电路零输入响应

Fig. 5-4 RC zero-input response

式中：

$$i = C \frac{du_C}{dt} \tag{5-4}$$

通解得：

$$u_C = U e^{-\frac{t}{RC}} \tag{5-5}$$

电流为：

$$i = C \frac{du_C}{dt} = -\frac{U}{R} e^{-\frac{t}{RC}} = -\frac{U}{R} e^{-\frac{t}{\tau}} \tag{5-6}$$

令：

$$\tau = RC \tag{5-7}$$

τ 称为 RC 电路的时间常数，单位为 s。时间常数反映了电路过渡过程的快慢。工程中，常认为经过 $3\tau \sim 5\tau$ 时间后，电容放电结束。

Note：

(1) The RC circuit zero-input response is the discharging process with no motivation from the power supply.

(2) $\tau = RC$ is the time constant, which determines the variation speed of the circuit transient process.

(3) For engineering, the discharging is finished after the time of $3\tau \sim 5\tau$.

Example 5-2：

In Fig. 5-5, $C = 4F$, $R_1 = R_2 = 20\Omega$ and $U_0 = 100V$. Find the discharging current I and the capacitor voltage u_C after the switching on at 60s.

Solution：

Take the switching off at $t = 0$,

Hence：

$$\tau = RC = \frac{R_1 R_2}{R_1 + R_2} C = \frac{20 \times 20}{20 + 20} \times 4 = 40s$$

$$u_C = U e^{-\frac{t}{\tau}} = 100 \times e^{-\frac{60}{40}} = 22.31V$$

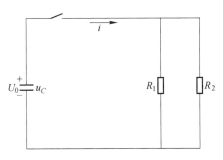

图 5-5 例 5-2

Fig. 5-5 Example 5-2

$$I = \frac{U}{R} = \frac{22.31}{10} = 2.231\text{A}$$

5.2.2 RC 电路的零状态响应（RC zero-state response）

RC 电路的零状态响应定义为电路的储能元件——电容的初始储能为零，通过外部电源对电容进行充电的过程。

如图 5-6 所示，换路前，开关未接通；$t = 0$ 时，开关接通，此时电源为电容充电。电容元件初始值为零。

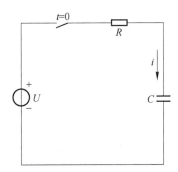

图 5-6 RC 电路零状态响应

Fig. 5-6 RC zero-state response

根据 KVL，可得：

$$RC \frac{du_C}{dt} + u_C = U \tag{5-8}$$

通解得：

$$u_C = -U e^{-\frac{t}{\tau}} + U = U(1 - e^{-\frac{t}{\tau}}) \tag{5-9}$$

其中：

$$\tau = RC \tag{5-10}$$

τ 表示充电的快慢。工程中，经常认为电路经过 $3\tau \sim 5\tau$ 时间后，电容充电结束。

Note：

(1) The RC circuit zero state response is under the condition of the zero initial energy value

of the capacitor and the circuit is only stimulated by the power source after charging.

(2) For engineering, the charging is finished after the time of $3\tau \sim 5\tau$.

Example 5-3:

In Fig. 5-7, the $u_C = 0V$ when the switch is off. Find the $i(0_+)$ and τ when the switch is turned on.

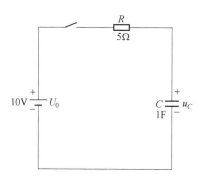

图 5-7 例 5-3

Fig. 5-7　Example 5-3

Solution:

When the switch is turned on,

$$i(0_+) = \frac{U_0}{R} = \frac{10}{5} = 2A$$

$$\tau = RC = 5 \times 1 = 5s$$

5.2.3 RC 电路的全响应 (RC complete response)

RC 电路的全响应定义为电源激励和电容元件的初始状态均不为零时电路的响应, 即零输入响应和零状态响应的叠加。

即

$$u_C = u_C(0_\infty) + [u_C(0_+) - u_C(0_\infty)] e^{-\frac{t}{RC}} \tag{5-11}$$

全响应=稳态分量+暂态分量

或者

$$u_C = u_C(0_+) e^{-\frac{t}{\tau}} + u_C(0_\infty)(1 - e^{-\frac{t}{\tau}}) \tag{5-12}$$

全响应=零输入响应+零状态响应

Note:

(1) RC circuit complete response is under the condition of the initial voltage value is not zero and the power source's voltage motivating the circuit is also not zero.

(2) The complete response is the superposition of the zero-input response and the zero-state response.

5.3 一阶 RL 电路的响应（First-order RL circuit）

RL 电路换路时，会产生零输入响应、零状态响应和全响应。

Note：

When the RL circuit switches, the response states include zero input response, zero state response and the complete response.

5.3.1 RL 电路的零输入响应（RL zero-input response）

如图 5-8 所示，换路前，开关位置为 1，电源对电感充电，$t=0$ 时，开关位置为 2，此时无电源提供能量，输入信号为零，电感将储有的能量释放。

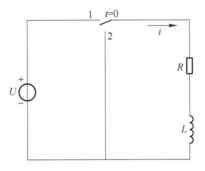

图 5-8 RL 零输入响应

Fig. 5-8 RL zero-input response

根据 KVL 得：

$$L\frac{\mathrm{d}i}{\mathrm{d}t} + Ri = 0 \tag{5-13}$$

通解为：

$$i(t) = \frac{U}{R}\mathrm{e}^{-\frac{t}{L/R}} = i(0_+)\mathrm{e}^{-\frac{t}{\tau}} \tag{5-14}$$

电压为：

$$u_L(t) = L\frac{\mathrm{d}i}{\mathrm{d}t} = -Ri(0_+)\mathrm{e}^{-\frac{t}{\tau}} \tag{5-15}$$

$$u_R(t) = Ri = Ri(0_+)\mathrm{e}^{-\frac{t}{\tau}} \tag{5-16}$$

其中：

$$\tau = \frac{L}{R} \tag{5-17}$$

τ 被称为 RL 电路的时间常数，单位为秒（s）。

Note：

(1) The RL circuit zero-input response is the discharging process with no motivations from the power supply.

(2) τ is the time constant, which determines the variation speed of the circuit transient process.

Example 5-4:

In Fig. 5-9, the circuit is steady with the switching off. Find the i_1, i_2, i_3 and u_L after the switch is turned on.

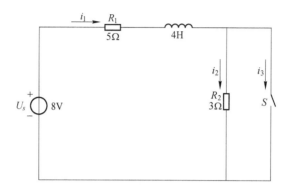

Fig. 5-9 Example 5-4

Solution:

Before the switching on,

$$i_1(0_-) = i_2(0_-) = \frac{U_s}{R_1 + R_2} = \frac{8}{5 + 3} = 1\text{A}$$

When the switch is turned on, according to the switching law:

$$i_L(0_+) = i_L(0_-) = i_1(0_+) = i_2(0_-) = 1\text{A}$$

With the state of switching on, no current flows through the resistor R_2.

So

$$i_2(0_+) = 0\text{A}$$
$$i_3(0_+) = i_1(0_+) - i_2(0_+) = 1\text{A}$$
$$U_s = R_1 i_1(0_+) + u_L(0_+)$$
$$u_L(0_+) = U_s - R_1 i_1(0_+) = 8 - 5 = 3\text{V}$$

5.3.2 RL 电路的零状态响应 (RL zero-state response)

如图 5-10 所示,电路换路前,$i_L(0_-) = 0$。

换路后,根据 KVL:

$$L\frac{\mathrm{d}i}{\mathrm{d}t} + Ri = U \tag{5-18}$$

通解为:

$$i = \frac{U}{R}\left(1 - \mathrm{e}^{-\frac{t}{L/R}}\right) = i(\infty)\left(1 - \mathrm{e}^{-\frac{t}{L/R}}\right) \tag{5-19}$$

令:

5.3 一阶 RL 电路的响应（First-order RL circuit）

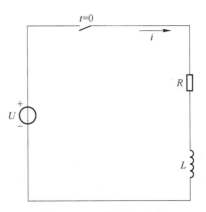

图 5-10 RL 零状态响应

Fig. 5-10 RL zero-state response

$$\tau = \frac{L}{R} \tag{5-20}$$

电压为：

$$u_R = U(1 - e^{-\frac{t}{\tau}}) \tag{5-21}$$

$$u_L = U e^{-\frac{t}{\tau}} \tag{5-22}$$

Note：

The RL circuit zero state response is under the condition of the zero initial energy value of the inductor.

5.3.3 RL 电路的全响应（RL complete response）

RL 电路的全响应定义为电源激励和电感元件的初始状态均不为零时电路的响应，如图 5-11 所示。

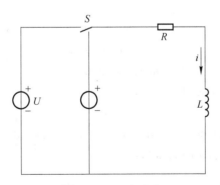

图 5-11 RL 全响应

Fig. 5-11 RL complete response

$$L\frac{di}{dt} + Ri = U \tag{5-23}$$

求解得 RL 电路的全响应为：

$$i = \frac{U}{R} + \left[i(0_+) - \frac{U}{R}\right]e^{-\frac{R}{L}t} \tag{5-24}$$

Note:

(1) RL circuit complete response is under the condition of the initial voltage value is not zero and the power source's voltage motivating the circuit is also not zero.

(2) RL circuit responses are coincident with the three element method.

5.4 一阶电路的三要素法（Three-element method of the first order linear circuit）

针对一阶电路的三种响应模式，可以根据电路中电量的初始值 $f(0_+)$、稳态值 $f(\infty)$ 和时间常数 τ 三个要素来求解。

其计算公式为：

$$f(t) = f(\infty) + [f(0_+) - f(\infty)]e^{-\frac{t}{\tau}} \tag{5-25}$$

where $f(t)$ ——任意瞬时电路中电压或电流；

$f(0_+)$ ——所求变量初始值；

$f(\infty)$ ——所求变量稳态值；

τ ——时间常数。

计算步骤：

——计算初始值：根据换路定律求解。

——计算稳态值：根据换路后的电路计算所求变量的稳态值（电容视为开路，电感视为短路）。

——计算时间常数：根据戴维南定理或者诺顿定理计算时间常数。对于电容电路，$\tau = RC$。

Note:

(1) Three elements are initial value, steady state values, and the time constant.

(2) $f(t)$ —any of the wanted current and voltage.

$f(0_+)$ —initial value of the wanted quantity.

$f(\infty)$ —corresponding steady state value.

τ—time constant.

(3) The calculation includes three steps: the initial value calculation, the steady state value calculation and the time constant calculation.

Example 5-5:

In Fig. 5-12, the circuit is steady at $t<0$. The switch is turned on at $t=0$. Find the $i_L(t)$ and $i_0(t)$ at $t \geq 0_+$.

Solution:

For the inductor:

$$i_L(0_+) = i_L(0_-) = 0A$$

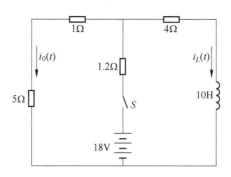

图 5-12 例 5-5

Fig. 5-12 Example 5-5

Find $i_L(\infty)$.

When $t \to \infty$, the inductor is as a short circuit, so

$$i_L(\infty) = 3\text{A}$$

Find τ

$$\tau = \frac{L}{R_{\text{eq}}} = \frac{10}{5} = 2\text{s}$$

$$i_L(t) = 3\left(1 - e^{-\frac{t}{2}}\right)\text{A} \qquad (t \geq 0_+)$$

$$i_0(t) = \frac{4i_L + 10\dfrac{\mathrm{d}i_L}{\mathrm{d}t}}{6} = 2 + 0.5e^{-\frac{t}{2}}\text{A} \qquad (t \geq 0_+)$$

5.5 练习题（Exercises）

5-1 In Fig. 5-13, $R = 2\Omega$, the internal resistor of the volt meter is 2.5kΩ, and the power voltage is 4V. Assuming the circuit is steady before switching, find the terminal voltage of the voltmeter at the moment of switching off.

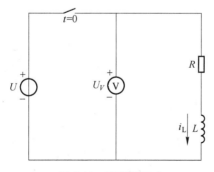

图 5-13 练习题 5-1

Fig. 5-13 Exercise 5-1

5-2 Find the time constant, τ, in Fig. 5-14.

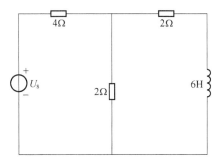

图 5-14 练习题 5-2
Fig. 5-14 Exercise 5-2

5-3 In Fig. 5-15, $U_s = 12V$, $R_1 = 4\Omega$, $R_2 = 2\Omega$, and the circuit is steady before switching on. Find the initial voltage value and the initial current after the switching on.

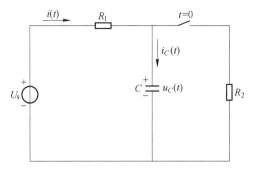

图 5-15 练习题 5-3
Fig. 5-15 Exercise 5-3

5-4 In Fig. 5-16, the circuit is in a steady state with switching on. Find the initial conditions of current.

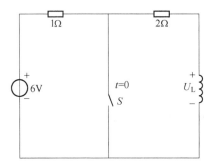

图 5-16 练习题 5-4
Fig. 5-16 Exercise 5-4

5-5 The circuit is steady before $t = 0$ in Fig. 5-17. Find $i(t)$ and $u(t)$ after the switch is turned on.

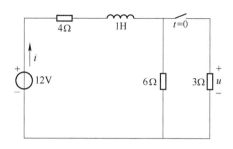

图 5-17 练习题 5-5
Fig. 5-17 Exercise 5-5

5-6 The circuit is steady before $t=0$ in Fig. 5-18. At $t=0$, the position of switch is from 1 to 2. Find $i_C(t)$, $u_C(t)$ and $i(t)$ at $t > 0$.

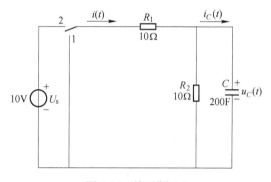

图 5-18 练习题 5-6
Fig. 5-18 Exercise 5-6

5-7 The circuit is steady before $t=0$ in Fig. 5-19. The switch is turned off at $t=0$. Find the value of the voltage $u_C(t)$ at $t = 2S$.

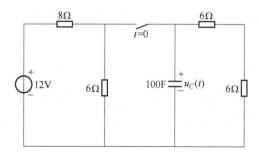

图 5-19 练习题 5-7
Fig. 5-19 Exercise 5-7

6 交流电 (Alternating Current)

目标 (Objectives)

When you have studied Chapter 6, you should
- have an understanding of alternating current (AC).
- be familiar with the alternating current parameters.
- be capable of analysis the AC with phasor representations.
- have an understanding of the effects of applying an alternating voltage across a resistor, an inductor or a capacitor.
- recognize and analyze the impedance in the series and parallel circuits.
- understand the term of power factor and the method to improve the power factor.

电压和电流随时间按规律变化称为交流电。

随时间变化按正弦规律变化的电压和电流称为正弦交流电压和正弦交流电流。

Note:

(1) An alternating current is an electrical current whose magnitude and direction vary cyclically.

(2) The sine wave is the most common wave in alternating current as sine AC.

6.1 正弦交流电及其特征参数 (AC and AC parameters)

6.1.1 正弦交流电 (AC)

正弦交流电路可以由若干独立的同频正弦交流电源组成，涉及的基本元件有电阻、电感和电容。直流电路中的支路电流法、叠加定理、戴维南定理和诺顿定理依然适用于正弦交流电路。

一个典型的正弦交流电如图6-1所示。

正弦交流电数学表达式：

$$u = U_m \sin(\omega t + \varphi_u) \tag{6-1}$$

$$i = I_m \sin(\omega t + \varphi_i) \tag{6-2}$$

where u, i——瞬时值 (instantaneous value);

U_m, I_m——最大值 (maximum value);

ω——角频率 (angular frequency);

φ_u,φ_i——初相角或者初相位(phase angle)。

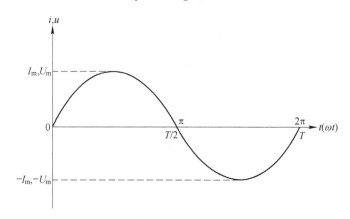

图 6-1 正弦交流电波形

Fig. 6-1 AC sinusoid waveform

6.1.2 特征参数(AC parameters)

正弦交流电的主要参数有最大值、瞬时值、有效值、周期、频率、角频率、相位与初相角等。

6.1.2.1 瞬时值(Instantaneous value)

定义：交流电在每一个时刻的值，叫作瞬时值。

符号：i，u。

Note：

Instantaneous value represents the magnitude of a waveform at instant time, which is denoted by lower-case symbols.

6.1.2.2 最大值(Maximum value)

定义：交流电瞬时值中的最大值，也叫作振幅值或者峰值。

符号：I_m，U_m。

Note：

The maximum value is the peak value in a period.

6.1.2.3 有效值(Effective value)

定义：交流电的有效值等于其瞬时值的平方在一个周期的平均值的算术平方根。

符号：I，U。

有效值与最大值的关系：

$$I = \frac{I_m}{\sqrt{2}} = 0.707 I_m \tag{6-3}$$

$$U = \frac{U_m}{\sqrt{2}} = 0.707 U_m \tag{6-4}$$

最大值与有效值的比称为波形系数，Fc。正弦交流电的波形系数为$\sqrt{2}$，三角形变量为$\sqrt{3}$。

Note：

(1) The root-mean-square (rms) value is also called effective value.

(2) The ratio of maximum value to the effective value of sine AC is $\sqrt{2}$.

Example 6-1：

Calculate the peak value when the rms value for a sinusoidal AC is 5A.

Solution：

According to the equation (6-3)：

$$I = \frac{I_m}{\sqrt{2}}$$

hence，

$$I_m = \sqrt{2} I = 1.414 \times 5 = 7.07 A$$

6.1.2.4 平均值（Average value）

定义：周期性交流量的波形曲线在半个周期内与横轴所围面积的平均值定义为交流量的平均值。

$$I_{ave} = \frac{2}{\pi} I_m = 0.637 I_m \tag{6-5}$$

Note：

The average value has a close relationship with half a cycle.

6.1.2.5 周期（Period）

定义：交流电完成一次循环变化的时间，称为周期。

符号：T。

单位：秒，s。

Note：

The period is the duration of one cycle.

6.1.2.6 频率（Frequency）

定义：单位时间内，交流电做周期性变化的次数。

符号：f。

单位：赫兹，Hz。

频率与周期的关系：

$$f = \frac{1}{T} \tag{6-6}$$

Note:

(1) Frequency is the number of cycles that occur in one second.

(2) The unit of frequency is hertz, Hz (Heinrich Rudolf Hertz).

Example 6-2:

An alternating voltage follows $u = 141.4\sin 377t$ V. Determine the values of (a) rms voltage, (b) frequency, and (c) the instantaneous voltage when $t = 3$ ms?

Solution:

According to the equation $u = 141.4\sin 377t$,

$$U_m = \sqrt{2}\,U = 141.4\text{V}$$
$$\omega = 2\pi f = 377 \text{rad/s}$$

Hence,

$$U = \frac{U_m}{\sqrt{2}} = \frac{141.4}{\sqrt{2}} = 100\text{V}$$

$$f = \frac{\omega}{2\pi} = \frac{377}{2 \times 3.14} = 60\text{Hz}$$

When $t = 3$ ms,

$$u = 141.4\sin(377 \times 3 \times 10^{-3}) = 127.8\text{V}$$

6.1.2.7 角频率 (Angular frequency)

定义:正弦交流电每秒内变化的电角度。

符号:ω。

单位:弧度/秒,rad/s。

角频率与频率的关系:

$$\omega = 2\pi f \tag{6-7}$$

6.1.2.8 相位与初相角 (Phase and initial phase angle)

正弦交流电压瞬时值和电流瞬时值的表达式中,$(\omega t + \varphi_u)$ 和 $(\omega t + \varphi_i)$ 称为电压正弦量和电流正弦量的相位(角)。

当 $t = 0$ 时,φ_u 和 φ_i 称为初相位(初相角)。

参考正弦量:

当电路中有多个相同频率正弦量同时存在时,可根据需要选择其中某一正弦量在由负向正变化通过零值的瞬间作为计时起点,即这个正弦量的初相就是零,称这个正弦量为参考正弦量。

同频率正弦量的相位差为其初相位之差。如:

$$u = U_m\sin(\omega t + \varphi_u) \tag{6-8}$$
$$i = I_m\sin(\omega t + \varphi_i) \tag{6-9}$$

则相位差为

$$j = (\omega t + \varphi_u) - (\omega t + \varphi_i) = \varphi_u - \varphi_i \tag{6-10}$$

当 $j = 0$ 时,称 u 与 i 同相(in phase),如图 6-2a 所示;
当 $j > 0$ 时,称 u 超前(lead)于 i,即 i 滞后(lag)于 u,如图 6-2b 所示;
当 $j = 90$ 时,称 u 与 i 正交(orthogonal in phase),波形如图 6-2c 所示;
当 $j = 180$ 时,称 u 与 i 反相(opposite in phase),波形如图 6-2d 所示。

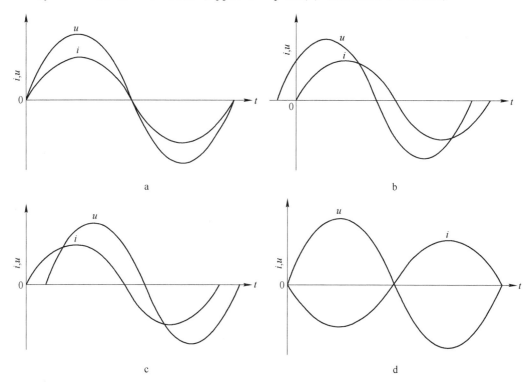

图 6-2 正弦量的相位关系
a—电压电流同相;b—电压超前电流;c—电压电流正交;d—电压电流反相
Fig. 6-2 Phase relationship of AC sinusoid
a—u and i in phase;b—u leads i;c—u and i orthogonal in phase;d—u and i opposite in phase

6.1.3 相量表示法 (Phasor representation method)

正弦交流电计算中,通常采用相量表示法进行电路分析计算,即使用复数(complex number)的运算方法。

复数如图 6-3 所示。

其中,复数 $\dot{A}$ 的模为:

$$r = \sqrt{a^2 + b^2} \tag{6-11}$$

辐角为

$$\varphi = \arctan \frac{b}{a} \tag{6-12}$$

正弦量相量的表示形式有四种,分别为代数形式、三角形式、指数形式和极坐标形式。

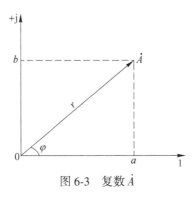

图 6-3 复数 $\dot{A}$

Fig. 6-3　Complex number $\dot{A}$

代数形式（algebra coordinate）：
$$\dot{A} = a + jb \tag{6-13}$$
三角形式（rectangular coordinate）：
$$\dot{A} = r\cos\varphi + jr\sin\varphi \tag{6-14}$$
指数形式（exponential coordinate）：
$$\dot{A} = re^{j\varphi} \tag{6-15}$$
极坐标形式（polar coordinate）：
$$\dot{A} = r\angle\varphi \tag{6-16}$$
四种相量的参数关系如图 6-3 所示，其中：
$$r = \sqrt{a^2 + b^2}$$
$$\varphi = \arctan\frac{b}{a} \tag{6-17}$$
$$a = r\cos\varphi \tag{6-18}$$
$$b = r\sin\varphi \tag{6-19}$$

对于 AC 交流电压 $u = U_m\sin(\omega t + \varphi_u)$ 和 AC 交流电流 $i = I_m\sin(\omega t + \varphi_i)$ 而言，其相量形式为：
$$\dot{U} = \frac{U_m}{\sqrt{2}}e^{j\varphi_u} = U\angle\varphi_u \tag{6-20}$$
$$\dot{I} = \frac{I_m}{\sqrt{2}}e^{j\varphi_i} = I\angle\varphi_i \tag{6-21}$$

Example 6-3：

Write down the phasor values from the following instantaneous values.
$$u = 311\sin(314t - 45°)\,\text{V}$$
$$i = 4.24\sin(314t + 60°)\,\text{A}$$

Solution：
$$\dot{U} = \frac{311}{\sqrt{2}}\angle -45° = 220\angle -45°\,\text{V}$$

$$\dot{I} = \frac{4.24}{\sqrt{2}} \angle 60° = 3\angle 60°\text{A}$$

6.2 交流电路中的电阻(Resistor in AC circuit)

6.2.1 电压和电流的关系(Relationship between voltage and current)

交流电路中,线性电阻的电压和电流参考方向如图6-4所示,且符合欧姆定律。

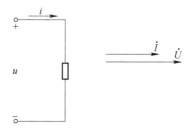

图6-4 电阻元件上的电压和电流

Fig. 6-4 Voltage and current of a resistor

已知:
$$i = \sqrt{2}I\sin\omega t \tag{6-22}$$

根据欧姆定律,
$$u = Ri = \sqrt{2}RI\sin\omega t = \sqrt{2}U\sin\omega t \tag{6-23}$$

则:
$$U = RI \tag{6-24}$$

已知电流相量为:
$$\dot{I} = I\angle 0° \tag{6-25}$$

则:
$$\dot{U} = U\angle 0° = RI\angle 0° = R\dot{I} \tag{6-26}$$

Note:

(1) The relationship between the voltage and current of a resistor in AC is determined by the Ohm's Law.

(2) The current and the voltage are in phase for resistors in AC.

6.2.2 功率(Power)

6.2.2.1 瞬时功率(Instantaneous power)

定义:交流电压和交流电流随着时间而变化,因此交流电通过电阻产生的功率也随之变化,某一时刻产生的功率称为瞬时功率。

符号:p。

单位：瓦特，W。

$$p = ui = U_m\sin\omega t \times I_m\sin\omega t$$
$$= U_m I_m \sin^2\omega t = UI(1 - \cos2\omega t) \tag{6-27}$$

Note:

(1) The instantaneous power is the product of instantaneous values of voltage and current.

(2) Instantaneous power changes with the time, and it is always greater than zero.

6.2.2.2 平均功率 (Average power)

定义：瞬时功率在一个周期内的平均值，称为平均功率，又称有功功率 (real power)。

符号：P。

单位：瓦特，W。

$$P = \frac{1}{T}\int_0^T p\,dt = \frac{1}{T}\int_0^T UI(1 - \cos2\omega t)\,dt = UI = I^2 R = \frac{U^2}{R} \tag{6-28}$$

有功功率等于电阻两端电压的有效值与通过电阻的电流有效值的乘积。

Note:

The average power of a resistor in AC is the product of effective values of voltage and current.

Example 6-4:

Find the instantaneous value of the current, effective values of the voltage and current, and the real power in Fig. 6-4, when R is 2Ω and the voltage is $u = 311\sin31.4t$ V.

Solution:

According to the Ohm's Law:

$$i = \frac{u}{R} = \frac{311}{2}\sin31.4t \text{ A}$$

$$I = \frac{I_m}{\sqrt{2}} = \frac{311}{2\sqrt{2}} = 110\text{A}$$

$$U = \frac{U_m}{\sqrt{2}} = \frac{311}{\sqrt{2}} = 220\text{V}$$

$$P = UI = 220 \times 110 = 24.2\text{kW}$$

6.3 交流电路中的电容器 (Capacitor in AC circuit)

6.3.1 电压和电流的关系 (Relationship between voltage and current)

交流电路中，电容元件的电压和电流参考方向如图 6-5 所示，电容电流的相位超前 (lead) 于电压 90°。

已知：

$$u = \sqrt{2}U\sin\omega t \tag{6-29}$$

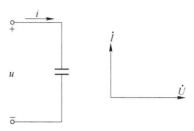

图 6-5 电容元件上的电压和电流

Fig. 6-5 Voltage and current of a capacitor

则:

$$i = C\frac{du}{dt} = C\frac{d}{dt}(\sqrt{2}U\sin\omega t) = \sqrt{2}\omega CU\cos\omega t = \sqrt{2}\omega CU\sin\left(\omega t + \frac{\pi}{2}\right) = \sqrt{2}I\sin\left(\omega t + \frac{\pi}{2}\right) \tag{6-30}$$

式中:

$$I = \omega CU \tag{6-31}$$

电流相量为:

$$\dot{I} = I\angle\varphi = \omega CU\angle 90° = \omega C\angle 90°U\angle 0° = j\omega C\dot{U} = j\frac{\dot{U}}{X_C} \tag{6-32}$$

或者:

$$\dot{U} = \frac{\dot{I}}{j\omega C} = -jX_C\dot{I} \tag{6-33}$$

式中:

$$X_C = \frac{1}{\omega C} = \frac{1}{2\pi fC} \tag{6-34}$$

X_C 被称为容抗 (capacitive reactance), 容抗表征电容元件对电流起阻碍作用的物理性质。交流电中, 电容元件有"隔直流通交流"的作用。

容抗单位: 欧姆, Ω。

Note:

(1) For the capacitor in AC, the phase of current is ahead of the voltage 90°.

(2) X_C represents the capacitive reactance. The unit is ohm, Ω.

(3) X_C is inverse proportional to the capacitance C and the frequency f.

(4) In AC, the capacitor has the function of isolating the direct current power.

6.3.2 功率 (Power)

6.3.2.1 瞬时功率 (Instantaneous power)

电容的瞬时功率 p 为:

$$p = ui = \sqrt{2}U\sqrt{2}I\sin\omega t\sin\left(\omega t + \frac{\pi}{2}\right) = 2UI\sin\omega t\cos\omega t = UI\sin 2\omega t \tag{6-35}$$

Note:

The instantaneous power is the product of the instantaneous voltage and instantaneous current.

6.3.2.2 平均功率 (Average power)

电容元件中，平均功率 P 为：

$$P = \frac{1}{T}\int_0^T p\,dt = \frac{1}{T}\int_0^T UI\sin2\omega t\,dt = 0 \tag{6-36}$$

交流电路中，电容元件与电源之间有能量互换，能量消耗为零。

Note:

(1) Capacitor does not cost any energy in AC. It just transfers the energy with the power source.

(2) The average power of a capacitor in AC is equal to zero.

6.3.2.3 无功功率 (Reactive power)

交流电路中，电容元件与电源的能量互换用无功功率 Q 衡量，无功功率是瞬时功率的最大值。

无功功率的单位为乏尔，var.

$$Q = UI = I^2 X_C = \frac{U^2}{X_C} \tag{6-37}$$

Note:

(1) The reactive power of a capacitor in AC is the maximum value of the instantaneous power.

(2) The unit of reactive power is var.

Example 6-5:

Find the X_C, instantaneous current i and reactive power Q, when a capacitor of $C = 31.8\mu F$ is connected to a power source $u = 380\sin\left(100\pi t + \frac{\pi}{4}\right)$ V.

Solution:

$$X_C = \frac{1}{2\pi f C} = \frac{1}{2 \times 3.14 \times 50 \times 31.8 \times 10^{-6}} = 100\Omega$$

$$I_m = \frac{U_m}{X_C} = \frac{380}{100} = 3.8A$$

As:

$$\varphi_i - \varphi_u = \frac{\pi}{2}, \quad \varphi_u = \frac{\pi}{4}$$

Hence,

$$\varphi_i = \frac{\pi}{2} + \frac{\pi}{4} = \frac{3}{4}\pi$$

$$i = 3.8\sin\left(100\pi t + \frac{3\pi}{4}\right) \text{A}$$

$$Q = UI = \frac{U_m I_m}{2} = \frac{380 \times 3.8}{2} = 722\text{var}$$

6.4 交流电路中的电感（Inductor in AC circuit）

6.4.1 电压和电流的关系（Relationship between voltage and current）

交流电路中，电容元件的电压和电流参考方向如图 6-6 所示，电感电流的相位滞后（lag）于电压 90°。

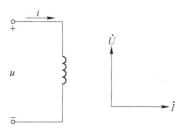

图 6-6 电感元件上的电压和电流

Fig. 6-6 Voltage and current of an inductor

已知：

$$i = \sqrt{2}I\sin\omega t \tag{6-38}$$

则：

$$u = L\frac{\mathrm{d}i}{\mathrm{d}t} = L\frac{\mathrm{d}}{\mathrm{d}t}(\sqrt{2}I\sin\omega t) = \sqrt{2}\omega IL\cos\omega t = \sqrt{2}\omega IL\sin\left(\omega t + \frac{\pi}{2}\right) = \sqrt{2}U\sin\left(\omega t + \frac{\pi}{2}\right) \tag{6-39}$$

式中：

$$U = \omega LI \tag{6-40}$$

电流相量为：

$$\dot{U} = U\angle\varphi = \omega LI\angle 90° = \omega L\angle 90° \angle 0° = \mathrm{j}\omega L\dot{I} \tag{6-41}$$

或者：

$$\dot{U} = \mathrm{j}X_L \dot{I} \tag{6-42}$$

式中：

$$X_L = \omega L = 2\pi fL \tag{6-43}$$

X_L 被称为感抗（inductive reactance），感抗是表征电感元件对电流起阻止能力大小的一个参数。感抗的大小随电流频率变化而变化，当电流频率为零，即直流时，感抗为零，即电感在直流稳态时相当于短路。

感抗的单位是欧姆，Ω。

Note:

(1) For the inductor in AC, the phase of current lags behind the voltage 90°.

(2) X_L represents the inductive reactance. The unit is ohm, Ω.

(3) X_L is proportional to the inductance L and the frequency f.

(4) The inductive coil has blocking effect for high frequency circuit. For DC circuit, the coil could be seen as a short circuit, when X_L is zero.

6.4.2 功率 (Power)

6.4.2.1 瞬时功率 (Instantaneous power)

电感的瞬时功率 p 为：

$$p = ui = \sqrt{2}U\sqrt{2}I\sin\omega t\sin\left(\omega t + \frac{\pi}{2}\right) = 2UI\sin\omega t\cos\omega t = UI\sin2\omega t \tag{6-44}$$

Note:

The instantaneous power is the product of the instantaneous voltage and instantaneous current.

6.4.2.2 平均功率 (Average power)

电感元件中，平均功率 P 为：

$$P = \frac{1}{T}\int_0^T p\,dt = \frac{1}{T}\int_0^T UI\sin2\omega t\,dt = 0 \tag{6-45}$$

交流电路中，电感元件与电源之间有能量互换，能量消耗为零。电感不消耗功率，为储能元件。

Note:

(1) There is no power consumption of the inductor in AC. It just transfers the energy with the power source.

(2) The average power of an inductor in AC is zero.

6.4.2.3 无功功率 (Reactive power)

交流电路中，电感元件与电源的能量互换用无功功率 Q 衡量，无功功率是瞬时功率的最大值。

$$Q = UI = I^2 X_L \tag{6-46}$$

Note:

(1) The reactive power of inductor in AC is the amplitude of the instantaneous power.

(2) The unit of reactive power is var.

Example 6-6:

An inductor of 1H is connected to an AC power source with the frequency of 5Hz and effective voltage of 10V. Find the current I.

Solution:

$$X_L = 2\pi f L = 2 \times 3.14 \times 5 \times 1 = 31.4\Omega$$

$$I = \frac{U}{X_L} = \frac{10}{31.4} = 0.32\text{A}$$

6.5 电阻、电容器和电感组成的电路（Resistor, capacitor and inductor in AC circuit）

6.5.1 阻抗（Impedance）

图 6-7 所示为电阻、电感和电容的串联交流电路图，称为 RLC 串联电路。

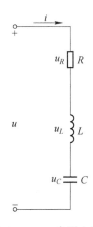

图 6-7 RLC 串联电路

Fig. 6-7 RLC series circuit

根据 KVL 定律，可得：

$$u = u_R + u_L + u_C \tag{6-47}$$

相量为：

$$\dot{U} = \dot{U}_R + \dot{U}_L + \dot{U}_C \tag{6-48}$$

设：

$$\dot{I} = I\angle 0° \tag{6-49}$$

则：

$$\dot{U}_R = \dot{I}R \tag{6-50}$$

$$\dot{U}_L = \dot{I}(\text{j}X_L) \tag{6-51}$$

$$\dot{U}_C = \dot{I}(-\text{j}X_C) \tag{6-52}$$

即：

$$\dot{U} = \dot{U}_R + \dot{U}_L + \dot{U}_C = \dot{I}R + \dot{I}(\text{j}X_L) + \dot{I}(-\text{j}X_C) = \dot{I}[R + \text{j}(X_L - X_C)] = \dot{I}Z \tag{6-53}$$

令：

$$Z = R + \text{j}(X_L - X_C) = R + \text{j}X \tag{6-54}$$

其中：
$$X = X_L - X_C \tag{6-55}$$

X 称为电抗（reactance）。

Z 称为阻抗（impedance）或者复阻抗（complex impedance），单位为欧姆，Ω。
$|Z|$ 称为阻抗模，表示为：

$$|Z| = \sqrt{R^2 + X^2} = \sqrt{R^2 + (X_L - X_C)^2} = \frac{U}{I} \tag{6-56}$$

阻抗三角形：阻抗模 $|Z|$、电阻 R 和电抗 X 构成的直角三角形称为阻抗三角形，如图 6-8 所示。

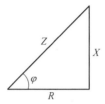

图 6-8 阻抗三角形

Fig. 6-8 Impedance rectangle

φ 为阻抗角（argument of impedance）：

$$\varphi = \arctan\left(\frac{X_L - X_C}{R}\right) = \varphi_u - \varphi_i \tag{6-57}$$

其中电感性电路，$\varphi > 0$，$X_L > X_C$。

电容性电路，$\varphi < 0$，$X_L < X_C$。

电阻性电路，$\varphi = 0$，$X_L = X_C$。

Note：

(1) φ is the argument of impedance, which is the phase difference of the voltage and the current.

(2) If $X_L > X_C$, the impedance argument $\varphi > 0$, the voltage is ahead of the current by the angle φ;

If $X_L < X_C$, the impedance argument $\varphi < 0$, the voltage lags behind the current by the angle φ;

If $X_L = X_C$, the impedance argument $\varphi = 0$, the voltage and the current are in phase, so the circuit is resistant.

Example 6-7：

In the circuit shown in Fig. 6-9, $R = 5\Omega$, $L = 31.8$mH, and the sinusoidal AC $U = 220$V, $f = 50$Hz. Determine the circuit current and the phase angle.

Solution：

According to equation (6-43)：

$$X_L = 2\pi fL = 2 \times 3.14 \times 50 \times 31.8 \times 10^{-3} = 10\Omega$$

According to equation (6-56)：

$$|Z| = \frac{U}{I} = \sqrt{R^2 + X_L^2} = \sqrt{5^2 + 10^2} = 11.2\Omega$$

$$I = \frac{U}{|Z|} = \frac{100}{11.2} = 8.9\text{A}$$

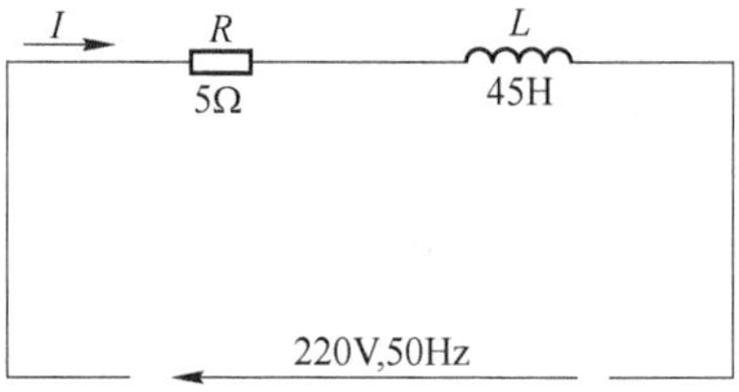

Fig. 6-9 Example 6-7

According to equation (6-57)

$$\varphi = \arctan\frac{X_L}{R} = \arctan\frac{10}{5} = 63.43°$$

Example 6-8:

In the circuit shown in Fig. 6-10, $I = 1\text{A}$, $C = 8\mu\text{F}$, and the sinusoidal AC $U = 220\text{V}$. (a) Find the voltage frequency, and (b) if the current is reduced to 0.5A, determine the resistance to be connected in series.

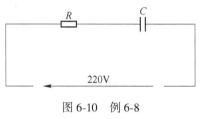

Fig. 6-10 Example 6-8

Solution:

For (a):

$$X_C = \frac{1}{2\pi fC} = \frac{U}{I} = \frac{220}{1} = 220\Omega$$

Hence,

$$f = \frac{1}{2\pi CX_C} = \frac{1}{2 \times 3.14 \times 8 \times 10^{-6} \times 220} = 90.5\text{Hz}$$

For (b), a resistor is connected in series as shown in Fig. 6-10:

$$|Z| = \sqrt{R^2 + X_C^2} = \frac{U}{I} = \frac{220}{0.5} = 440\Omega$$

Hence,

$$R = \sqrt{|Z|^2 - X_C^2} = \sqrt{440^2 - 220^2} = 381\Omega$$

6.5.2 阻抗的串联和并联电路（Series and parallel circuit of impedance）

正弦交流电路中，欧姆定律和基尔霍夫定律等均适用于阻抗的串联和并联电路，但计算时需要采用复数形式。

6.5.2.1 阻抗的串联（Impedance series circuit）

如图 6-7 所示，根据串联电路特性，电路外加电压为 u，通过电路各部件的电流 i 相等。

$$Z = \sum_{i=1}^{n} Z_i = Z_R + Z_L + Z_C \tag{6-58}$$

$$Z_R = R \tag{6-59}$$

$$Z_C = \frac{1}{j\omega C} \tag{6-60}$$

$$Z_L = j\omega L \tag{6-61}$$

根据欧姆定律，可得：

$$\dot{U} = \dot{I}Z = \dot{I}(Z_R + Z_L + Z_C) \tag{6-62}$$

RLC 串联交流电路的瞬时功率为：

$$p = ui = \sqrt{2}U\sqrt{2}I\sin\omega t \sin(\omega t + \varphi) = UI\cos\varphi - UI\cos(2\omega t + \varphi) \tag{6-63}$$

平均功率为：

$$P = \frac{1}{T}\int_0^T p\,dt = \frac{1}{T}\int_0^T [UI\cos\varphi - UI\cos(2\omega t + \varphi)]\,dt = UI\cos\varphi \tag{6-64}$$

其中：

$\cos\varphi$ 称为功率因素（power factor），$0 \leq \cos\varphi \leq 1$。

无功功率为：

$$Q = U_L I - U_C I = (U_L - U_C)I = I^2(X_L - X_C) = UI\sin\varphi \tag{6-65}$$

视在功率（apparent power）为：

$$S = UI = |Z|I^2 = \sqrt{P^2 + Q^2} \tag{6-66}$$

视在功率的单位是伏·安，V·A。

Note：

(1) The identical impedance is equal to the sum of the series impedances.

(2) $\cos\varphi$ is the power factor.

6.5.2.2 阻抗的并联（Impedance parallel circuit）

图 6-11 所示为几个阻抗的并联交流电路图。根据并联电路特性，通过电路各部件的外加电压 u 相等。

阻抗 Z 的关系为：

$$\frac{1}{Z} = \sum_{i=1}^{n} \frac{1}{Z_i} = \frac{1}{Z_R} + \frac{1}{Z_L} + \frac{1}{Z_C} \tag{6-67}$$

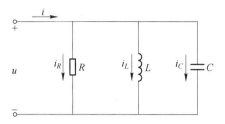

图 6-11 阻抗并联电路

Fig. 6-11 Parallel circuit of impedance

根据 KCL 定律,可得:

$$\dot{I} = \dot{I}_R + \dot{I}_L + \dot{I}_C = \frac{\dot{U}}{Z_R} + \frac{\dot{U}}{Z_L} + \frac{\dot{U}}{Z_C} \tag{6-68}$$

Note:

(1) The reciprocal of the identical impedance is equal to the sum of each parallel impedance reciprocal.

(2) The total current is the sum of the branch currents.

(3) The power and the impedance are not sine functions and could not be presented by phasors.

Example 6-9:

In a parallel circuit, a resistor of 110Ω and a capacitor of 63.5μF are connected to a 220V, 50Hz power source. Find the I_R, I_L, I and impedance Z.

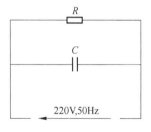

图 6-12 例 6-9

Fig. 6-12 Example 6-9

Solution:

$$I_R = \frac{U}{R} = \frac{220}{110} = 2\text{A}$$

$$X_C = \frac{1}{2\pi fC} = \frac{1}{2 \times 3.14 \times 50 \times 63.5 \times 10^{-6}} = 50\Omega$$

$$I_C = \frac{U}{X_C} = \frac{220}{50} = 4.4\text{A}$$

Hence:

$$I = \sqrt{I_R^2 + I_C^2} = \sqrt{2^2 + 4.4^2} = 4.8\text{A}$$

$$Z = \frac{U}{I} = \frac{220}{4.8} = 45.8\Omega$$

Example 6-10:

In Fig. 6-13, $U=220\text{V}$, $f=50\text{Hz}$, $R=50\Omega$, $L=2\text{H}$, and $C=50\mu\text{F}$. Find the current in each branch and the total current.

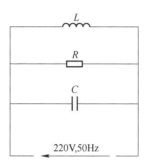

图 6-13 例 6-10

Fig. 6-13 Example 6-10

Solution:

$$I_R = \frac{U}{R} = \frac{220}{50} = 4.4\text{A}$$

$$I_C = \frac{U}{X_C} = 2\pi fCU = 2 \times 3.14 \times 50 \times 50 \times 10^{-6} \times 220 = 3.45\text{A}$$

$$I_L = \frac{U}{X_L} = \frac{U}{2\pi fL} = \frac{220}{2 \times 3.14 \times 50 \times 0.15} = 0.35\text{A}$$

$$I = \sqrt{I_R^2 + (I_C - I_L)^2} = \sqrt{4.4^2 + (3.45 - 0.35)^2} = 5.38\text{A}$$

6.6 谐振电路（Resonance circuit）

定义：含有电容和电感元件的正弦电路中，在特定条件下，电路端口电压与电流相位相同时，称为谐振电路。

电路实现谐振时，整个电路负载呈电阻性（resistive）。

6.6.1 串联谐振（Series resonance）

如图 6-14 所示 RLC 串联电路中，复阻抗为：

$$Z = R + \text{j}(X_L - X_C) = R + \text{j}\left(\omega L - \frac{1}{\omega C}\right) \qquad (6-69)$$

当 $X_L = X_C$ 时，$|Z| = R$，此时电压与电流同向，呈电阻性，发生谐振。

谐振时，电路阻抗值最小，电压一定时，电流最大。

谐振时，电路的谐振角频率（resonance angular frequency）ω_0 和谐振频率（resonance frequency）f_0 为：

$$\omega_0 = \frac{1}{\sqrt{LC}} \tag{6-70}$$

$$f_0 = \frac{1}{2\pi\sqrt{LC}} \tag{6-71}$$

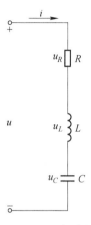

图 6-14　RLC 串联电路

Fig. 6-14　RLC series circuit

串联谐振时的感抗和容抗被称为电路的特性阻抗（characteristic impedance），用 ρ 表示：

$$\rho = \omega_0 L = \frac{1}{\omega_0 C} = \frac{\sqrt{LC}}{C} = \sqrt{\frac{L}{C}} \tag{6-72}$$

谐振时，电感和电容的电压大小相等，相位相反，且大小为电源电压 U_s 的 Q 倍。Q 被称为电路的品质因数（quality factor）。

$$Q = \frac{U_L}{U_s} = \frac{I\omega_0 L}{IR} = \frac{\omega_0 L}{R} = \frac{\rho}{R} \tag{6-73}$$

电流随频率变化的曲线称为电流谐振曲线（resonance curve），如图 6-15 所示。

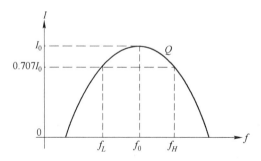

图 6-15　电流-频率变化曲线

Fig. 6-15　Current-frequency curve

在谐振点 f_0，电路的电流最大，$I = I_0$，此外，$I < I_0$。

6.6 谐振电路 (Resonance circuit)

当电路电流为谐振电流的 $\frac{1}{\sqrt{2}}$ 时，谐振曲线上两个对应点的频率 f_L 和 f_H 之间的范围，称为电路的通频带 (passband)，f_{BW}。

通频带的大小与品质因数 Q 有关。通频带与品质因数的关系为：

$$f_{BW} = f_H - f_L = \frac{f_0}{Q} \tag{6-74}$$

式中，f_H 和 f_L 称为上下截止频率。

Note:

(1) Electrical resonance occurs in an electric circuit at a particular resonant frequency when the imaginary parts of impedances or admittances of circuit elements cancel each other.

(2) At resonance, the series impedance of the two elements is the minimum value.

Example 6-11:

Find the capacitance, C, required to give series resonance and the capacitor voltage in a series RLC circuit in the condition of $R = 5\Omega$, $L = 250\text{mH}$, $U_{AC} = 100\text{V}$, and $f = 100\text{Hz}$.

Solution:

At the resonance

$$X_C = X_L = 2\pi f L = 2 \times 3.14 \times 100 \times 0.25 = 157.1\Omega$$

$$C = \frac{1}{2\pi f X_C} = \frac{1}{2 \times 3.14 \times 100 \times 157.1} = 10.15\mu\text{F}$$

$$I = \frac{U}{Z} = \frac{100\text{V}}{5\Omega} = 20\text{A}$$

$$U_C = IX_C = 20\text{A} \times 157.1\Omega = 3014.2\text{V}$$

Example 6-12:

The bandwidth of a series resonant circuit is 500Hz. (a) If the resonant frequency is 6000Hz, what is the value of Q? (b) If $R = 10\Omega$, what is the value of the inductive reactance at resonance? (c) Calculate the inductance and capacitance of the circuit.

Solution:

For (a), according to equation (6-14):

$$f_{BW} = \frac{f_0}{Q}$$

Hence:

$$Q = \frac{f_0}{f_{BW}} = \frac{6000}{500} = 12$$

For (b),

$$Q = \frac{X_L}{R}$$

Hence,

$$X_L = QR = 2\pi f_0 L = 12 \times 10 = 120\Omega$$

$$L = \frac{X_L}{2\pi f_0} = \frac{120}{2 \times 3.14 \times 6000} = 3.18 \text{mH}$$

For (c),
$$|X_L| = |X_C| = 120\Omega$$
$$X_C = \frac{1}{2\pi f_0 C} = 120\Omega$$

Hence,
$$C = \frac{1}{2\pi f_0 X_C} = \frac{1}{2 \times 3.14 \times 120 \times 6000} = 0.22\mu\text{F}$$

6.6.2 并联谐振 (Parallel resonance)

RLC 并联电路如图 6-16 所示，在关联方向下，如果电路的总电流与端电压同相，即 $R \ll \omega L$，则会发生并联谐振。

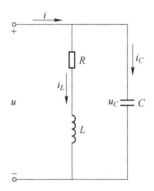

图 6-16 RLC 并联电路

Fig. 6-16 RLC parallel circuit

并联谐振时，电路的谐振角频率 ω_0 和谐振频率 f_0 如下：

$$\omega_0 = \frac{1}{\sqrt{LC}} \tag{6-75}$$

$$f_0 = \frac{1}{2\pi\sqrt{LC}} \tag{6-76}$$

并联谐振电路的阻抗为最大值，且具有纯电阻性。

$$Z = \frac{R^2 + (2\pi f_0 L)^2}{R} = \frac{L}{RC} \tag{6-77}$$

并联谐振电路总电流最小，与端电压同相。

并联谐振电路的品质因数为谐振时各支路电流与总电流的比值 Q，

$$Q = \frac{I_L}{I_0} = \frac{I_C}{I_0} = \frac{\omega_0 L}{R} = \frac{1}{\omega_0 CR} \tag{6-78}$$

Note:

The quality factor of parallel resonance is the ratio of branch current to the total current.

Example 6-13:

In Fig. 6-17, a resistor of 1kΩ and an inductor of 0.15H are connected in series. A variable capacitor is connected in parallel with the resistor and inductor. The power source is an AC supply with 2V and 10kHz. Determine (a) the capacitance when the supply current is a minimum, (b) the effective impedance at resonance, and (c) the circuit current.

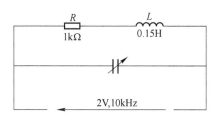

图 6-17 例 6-13
Fig. 6-17 Example 6-13

Solution:

For (a),
$$f_0 = \frac{1}{2\pi\sqrt{LC}}$$

Hence,
$$C = \frac{1}{4\pi^2 f_0^2 L} = \frac{1}{4 \times 3.14^2 \times 0.15 \times 10^8} = 1.69 \times 10^{-9} \text{F}$$

For (b),
$$Z = \frac{L}{CR} = \frac{0.15}{1.69 \times 10^{-9} \times 1000} = 89 \text{k}\Omega$$

For (c),
$$I = \frac{U}{Z} = \frac{2}{89 \times 10^3} = 22.5 \mu\text{A}$$

6.7 功率因数的提高（Improvement of power factor）

6.7.1 功率因数（Power factor）

定义：用电负载的有功功率与视在功率的比值称为功率因数，符号为 λ。

$$\lambda = \frac{P}{S} = \cos\varphi \tag{6-79}$$

cosφ 介于 0 和 1 之间。电阻性负载时，功率因数为 1。

在实际工程中，大多数负载都呈感性。如果功率因数低于 1，电路中会出现无功功率，发电设备的容量利用不充分，且会导致输电线路的损耗和压降增加。

Note:

(1) Power factor of an AC electrical power system is the ratio of the real power to the apparent power in the circuit.

(2) The value of power factor is in the closed interval of zero to one.

Example 6-14:

In Fig. 6-18, an inductor coil is connected to a source of 220V at 50Hz and the current flowing through is 8A. The coil dissipates 720W. Find the coil resistance, the coil inductance and the power factor.

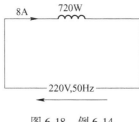

图 6-18　例 6-14

Fig. 6-18　Example 6-14

Solution:

$$Z = \frac{U}{I} = \frac{220}{8} = 27.5\Omega$$

$$R = \frac{P}{I^2} = \frac{750}{64} = 11.7\Omega$$

$$X_L = \sqrt{Z^2 - R^2} = \sqrt{27.5^2 - 11.7^2} = 24.9\Omega$$

$$L = \frac{X_L}{2\pi f} = \frac{24.9}{2 \times 3.14 \times 50} = 0.08H$$

$$\lambda = \cos\varphi = \frac{P}{S} = \frac{P}{UI} = \frac{750}{220 \times 8} = 0.43$$

6.7.2　功率因数的提高 (Improvement of power factor)

为了提高功率因数，通常在负载两端并联一个合适的电容器进行功率补偿。

感性负载并联电容后，由于电容元件的无功功率与电感元件的无功功率相互补偿，从而减少了电源供给的无功功率，但电源提供的有功功率并没有改变，从而提高了整个电路的功率因数。

要使电路的功率因数由原来的 $\cos\varphi_1$ 提高到 $\cos\varphi_2$，需并联的电容器的电容量为：

$$C = \frac{P}{\omega U^2}(\tan\varphi_1 - \tan\varphi_2) \tag{6-80}$$

where　P——感性负载的有功功率 (real power), W;

　　　ω——电源角频率 (angular frequency), rad/s;

　　　U——电源电压的有效值 (effective voltage), V。

Note:

(1) A capacitor is connected in parallel to the load in circuit to improve the power factor.

(2) The required capacitance is related to the real power of the load, angular frequency of power source, effective voltage of power source, and the original and target arguments.

Example 6-15:

How to improve the power factor to 0.9 in Fig. 6-16, when $u = 220$V, $f = 50$Hz, original power factor is 0.7, and the power is 4kW?

Solution:

A capacitor can be connected in the circuit in parallel to improve the power factor.

Suppose the original power factor angle is φ_1 and the power factor angle after adding the capacitor is φ_2.

So:

$$C = \frac{P}{\omega U^2}(\tan\varphi_1 - \tan\varphi_2) = \frac{4000}{2 \times 3.14 \times 50 \times 220^2}(\tan 0.7 - \tan 0.9) = 141\mu F$$

6.8 练习题 (Exercises)

6-1 Find the maximum value (I_m), effective value (I), frequency (f), and the initial phase (ψ) in Fig. 6-19.

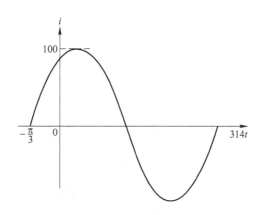

图 6-19　练习题 6-1

Fig. 6-19　Exercise 6-1

6-2 In an AC system, the voltage is $u = 311\sin(314t - 30°)$V. Find the period T, angular frequency ω, frequency f, maximum value U_m, phase, and initial phase. If $t = \dfrac{T}{2}$, find the instantaneous value of voltage.

6-3 A coil connected to a 230V, 50Hz sinusoidal supply takes a current of 10A at a phase angle of 30°. Calculate the resistance, inductance and the power taken by the coil.

6-4 Find the complex impedance Z of the network shown in Fig. 6-20.

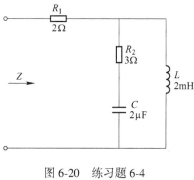

图 6-20　练习题 6-4
Fig. 6-20　Exercise 6-4

6-5　Find the current $i(t)$, $i_L(t)$ and $i_C(t)$ in Fig. 6-21. The source voltage is $u(t) = 10\sqrt{2}\sin(100t + 30°)$.

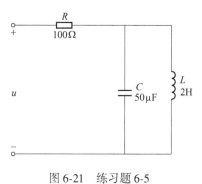

图 6-21　练习题 6-5
Fig. 6-21　Exercise 6-5

6-6　Find the power P, Q and S supplied by the current source $\dot{I} = 16\angle 0°$ in Fig. 6-22.

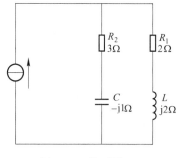

图 6-22　练习题 6-6
Fig. 6-22　Exercise 6-6

6-7　When $\omega = \dfrac{1}{\sqrt{LC}}$, decide whether the circuits are short circuit in Fig. 6-23.

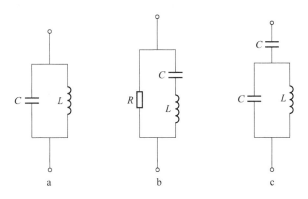

图 6-23　练习题 6-7

Fig. 6-23　Exercise 6-7

6-8　A coil having a resistance of 8Ω and an inductance of 0.2H is connected across a 200V, 50Hz supply. Find (a) the reactance and the impedance of the coil, (b) the current, and (c) the phase difference between the current and the applied voltage.

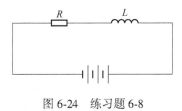

图 6-24　练习题 6-8

Fig. 6-24　Exercise 6-8

6-9　A resistor ($P = 150\text{W}$ and $U = 20\text{V}$) is connected in series with a capacitor across a 220V, 50Hz power supply. Find the capacitance required and the phase angle between the current and the supply voltage.

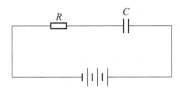

图 6-25　练习题 6-9

Fig. 6-25　Exercise 6-9

6-10　In a circuit, a resistor having a resistance of 12Ω, an inductor having an inductance of 0.15H and a capacitor of capacitance of 100mF are connected in series across a 100V, 50Hz power supply. Determine the impedance, the current, the voltages across R, L and C, and the phase difference between the current and the supply voltage.

6-11　A non-inductive resistor is connected in series with a coil across a 230V, 50Hz supply. The current is 1.8A and the potential differences across the resistor and the coil are 80V and

170V respectively. Calculate the inductance and the resistance of the coil, and the phase difference between the current and the supply voltage.

6-12 A circuit, having a resistance of 4Ω and inductance of 0.5H and a variable capacitance in series, is connected across a 100V, 50Hz supply. Determine (a) the capacitance to give resonance, (b) the voltages across the inductance and the capacitance, and (c) the Q factor of the circuit.

6-13 A coil of resistance 5Ω and inductance of 1H is connected in series with a 0.1F capacitor. The circuit is connected to a 5V variable frequency supply. Calculate the frequency at which resonance occurs, the voltages across the coil and the capacitor at this frequency and the Q factor of the circuit.

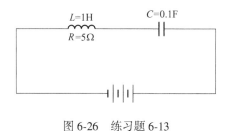

图 6-26 练习题 6-13
Fig. 6-26 Exercise 6-13

6-14 For the series resonant circuit with a source supply of 20V, calculate I at resonance with $R=4\Omega$, $X_L=j20\Omega$, and $X_C=-j20\Omega$. What are the voltages across the three series components, R, L and C? Calculate Q. If the resonant frequency is 6000Hz, find the bandwidth. What are the half-power frequencies, and what is the power dissipated in the circuit at the two frequencies?

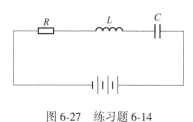

图 6-27 练习题 6-14
Fig. 6-27 Exercise 6-14

6-15 A RLC series circuit is connected to an AC power. The power source is $u = 100\sqrt{2}\sin 1000t$, and the quality factor is $Q=10$. Find the u_R, u_C and u_L when the circuit is at the resonance state.

6-16 A coil, of resistance R and inductance L, is connected in series with a capacitor C across a variable-frequency source. The voltage is maintained constant at 300mV and the frequency is varied until a maximum current of 5mA flows through the circuit at 6kHz. If, under these conditions, the Q factor of the circuit is 105, calculate: (a) the voltage across the capacitor; and (b) the values of R, L and C.

6-17 Fig. 6-28 represents a RLC parallel circuit. $I=5\text{A}$, $I_R=3\text{A}$, $I_C=3\text{A}$. Find I_L.

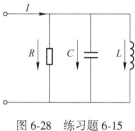

图 6-28 练习题 6-15
Fig. 6-28 Exercise 6-15

6-18 An inductive load is connected to a power source with $u=220\text{V}$, $f=50\text{Hz}$, the power factor of 0.6, and the consumed power of 4kW. How to improve the power factor to 0.9?

6-19 Find the $\dot{U}$, $\dot{I}$ and power factor $\cos\varphi$ in Fig. 6-29, when $\dot{U}_C=100\angle 0°$ in AC circuit.

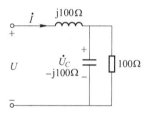

图 6-29 练习题 6-17
Fig. 6-29 Exercise 6-17

6-20 A load with $P=40\text{W}$ is connected to an AC power source with $U=220\text{V}$ and $f=50\text{Hz}$. The power factor is $\cos\varphi=0.5$, and the current is $I=0.367\text{A}$. Find the reactive power, Q. If the power factor increases to 0.9, find the total current, reactive power, and the capacitance required.

7 三相交流电（Three-phase Sinusoidal AC Circuit）

目标（Objectives）

In Chapter Seven, you should
- be familiar with balanced three-phase voltage.
- be capable of analyzing the wye and delta source connections.
- have an understanding of the three-phase loads.
- recognize and analyze the wye-wye circuit and delta-delta circuit.
- understand the three-phase power.

在电力系统中，电能的产生、传送、分配和使用几乎都采用三相交流电。本书主要讨论三相正弦交流电路。

7.1 对称三相交流电源（Balanced three-phase voltage source）

三个幅值相等、角频率相同、初相互差为120°的电动势，称为对称三相电动势或对称三相电源。

Three-phase circuit contains three voltage sources. If the three sinusoidal voltages have the same frequency and magnitude, and their phase angles are 120° apart with the other two, these three voltages are called balanced three-phase voltage source.

7.1.1 对称三相交流电压（Balanced three-phase voltage）

（1）瞬时值表达式（instantaneous value）

$$u_1 = \sqrt{2}U\sin(\omega t)$$
$$u_2 = \sqrt{2}U\sin(\omega t - 120°) \tag{7-1}$$
$$u_3 = \sqrt{2}U\sin(\omega t + 120°)$$

（2）相量表示（phasor）

$$\dot{U}_1 = U\angle 0°$$
$$\dot{U}_2 = U\angle -120° \tag{7-2}$$
$$\dot{U}_3 = U\angle 120°$$

(3) 波形图（oscillogram）（图 7-1）

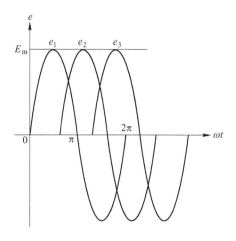

图 7-1 波形图

Fig. 7-1 Waveform diagram

(4) 相量图（phasor diagram）（图 7-2）

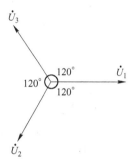

图 7-2 向量图

Fig. 7-2 Phase diagram

结论：

(1) 对称三相电源中，3 个正弦量的瞬时值之和为零。

(2) 对称三相电源中，3 个正弦量的相量和为零。

(3) 相序：对称三相电源到达振幅值（或零值）的先后次序。

　　　　A→B→C→A 称顺序（正序），

　　　　A→C→B→A 称逆序（反序）。

(4) 为使电力系统能够安全可靠运行，通常统一规定技术标准，一般在配电盘上用黄色标出 A 相，用绿色标出 B 相，用红色标出 C 相。

Note：

(1) The three-phase circuit is made of three-phase power source and three-phase load.

(2) Three-phase balanced sine voltages are with the same frequencies, same amplitudes and same phase differences (120°).

(3) The sum of the instantaneous value or phasor of the three-phase symmetric sine voltage is zero.

(4) The sequence of the existence of positive amplitude is called phase sequence. Positive phase sequence is A→B→C→A. If the B and C are interchanged, it is the negative phase sequence, A→C→B→A.

7.1.2 三相交流电源星形连接 (Three-phase source wye connection)

将发电机尾端 X、Y、Z 连接在一点，首端 A、B、C 分别与负载相连的方式叫作星形（Y）连接。如图 7-3 所示。

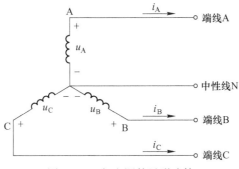

图 7-3 三相电源的星形连接

Fig. 7-3 Three-phase source wye connection

术语（Terms）：

中点或零点（neutral point）：三个尾端的公共连接点，用 N 表示。

中线或零线（null line）：中点引出的线。

端线或相线（phase line）：首端引出的三根线，俗称火线。

相电压（phase voltage）：相线与中线之间的电压（每相绕组首端与尾端的电压），U_p，复数用 $\dot{U}_A$、$\dot{U}_B$、$\dot{U}_C$ 表示。

线电压（line voltage）：相线与相线之间的电压（两相首端之间的电压），U_l，复数用 $\dot{U}_{AB}$、$\dot{U}_{BC}$、$\dot{U}_{CA}$ 表示。

相电流（phase current）：三相电源中流过每相负载的电流为相电流，I_p，复数用 $\dot{I}_{AB}$，$\dot{I}_{BC}$，$\dot{I}_{CA}$ 表示。

线电流（line current）：三相电源中每根导线中的电流为线电流，I_l，复数用 $\dot{I}_A$、$\dot{I}_B$、$\dot{I}_C$ 表示。

规定电压方向：规定电动势的方向为从绕组的尾端指向首端。

根据 KVL 定律和相量图可得三相对称电源星形连接时线电压和相电压有效值的通用关系：

$$\dot{U}_{AB} = \sqrt{3}\dot{U}_A \angle 30°$$
$$\dot{U}_{BC} = \sqrt{3}\dot{U}_B \angle 30° \quad (7\text{-}3)$$
$$\dot{U}_{CA} = \sqrt{3}\dot{U}_C \angle 30°$$

即线电压的有效值是相电压的有效值的$\sqrt{3}$倍，且相位超前对应的相电压30°。

$$U_l = \sqrt{3}\, U_p \tag{7-4}$$

星形连接的线电流与相电流相等，即

$$I_l = I_p \tag{7-5}$$

7.1.3 三相交流电源三角形连接（Three-phase source delta connection）

如果将三相电源的每相绕组首尾依次连接，称为三相电源的三角形（△）接法，见图7-4。

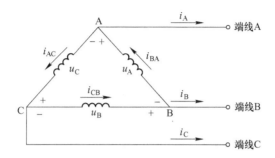

图7-4 三相电源的三角形连接

Fig. 7-4 Three-phase source delta connection

电源△连接时线电压等于相电压，即

$$U_l = U_p \tag{7-6}$$

线电流与相电流的关系为：

$$\begin{aligned}\dot{I}_A &= \dot{I}_{AB} - \dot{I}_{CA} = \sqrt{3}\,\dot{I}_{AB}\angle 30°\\ \dot{I}_B &= \dot{I}_{BC} - \dot{I}_{AB} = \sqrt{3}\,\dot{I}_{BC}\angle 30°\\ \dot{I}_C &= \dot{I}_{CA} - \dot{I}_{BC} = \sqrt{3}\,\dot{I}_{CA}\angle 30°\end{aligned} \tag{7-7}$$

即线电流是相电流的$\sqrt{3}$倍：

$$I_l = \sqrt{3}\, I_p \tag{7-8}$$

7.2 三相负载（Three-phase load）

7.2.1 星形连接（Wye-wye circuit）

三相交流电路星形连接如图7-5所示。

各相负载的相电压等于电源相电压：

$$\begin{aligned}\dot{U}'_A &= \dot{U}_A = U\angle 0°\\ \dot{U}'_B &= \dot{U}_B = U\angle -120°\\ \dot{U}'_C &= \dot{U}_C = U\angle 120°\end{aligned} \tag{7-9}$$

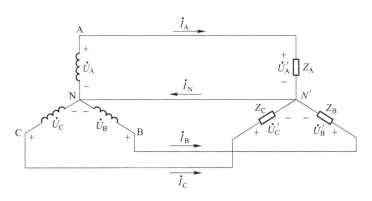

图 7-5 星形连接

Fig. 7-5 Wye-wye circuit

电路中的线电压等于负载相电压的 $\sqrt{3}$ 倍，即

$$\dot{U}_l = \sqrt{3}\,\dot{U}_p \tag{7-10}$$

线电流（phase current）：流过每根相线的电流，用 $\dot{I}_l$ 表示。

相电流（line current）：流过每相电源（负载）的电流，用 $\dot{I}_p$ 表示。

线电流与相电流的关系：

$$\dot{I}_l = \dot{I}_p \tag{7-11}$$

各相负载中通过的电流分别为 $\dot{I}_A$、$\dot{I}_B$、$\dot{I}_C$：

$$\dot{I}_A = \frac{\dot{U}_A}{Z_A} \tag{7-12}$$

$$\dot{I}_B = \frac{\dot{U}_B}{Z_B} \tag{7-13}$$

$$\dot{I}_C = \frac{\dot{U}_C}{Z_C} \tag{7-14}$$

中线电流（neutral current）：流过中线的电流，用 $\dot{I}_N$ 表示：

$$\dot{I}_N = \dot{I}_A + \dot{I}_B + \dot{I}_C \tag{7-15}$$

可见，在三相电源对称、三相 Y 接负载也对称的情况下，三相负载电流也是对称的，此时中线电流为零。

Note：

(1) If both of the sources and loads are balanced, there is no current flowing through the neural line.

(2) The relationship between the line voltage and phase voltage is $\dot{U}_l = \sqrt{3}\,\dot{U}_p$.

Example 7-1：

For a wye connected three-phase balanced load, the equivalent impedance is $Z = (6 +$

j8)Ω, and $u_{12} = 380\sqrt{2}\sin(\omega t + 30°)$ V. Find the phase currents.

Solution:

$$u_{12} = 380\sqrt{2}\sin(\omega t + 30°) \text{ V}$$

$$\dot{U}_{12} = 380\angle 30°$$

So

$$\dot{U}_1 = 220\angle 0°$$

$$\dot{I}_1 = \frac{\dot{U}_1}{Z_1} = \frac{220\angle 0°}{6 + j8} = \frac{220\angle 0°}{10\angle 53°} = 22\angle -53° \text{ A}$$

As the load is symmetric, so

$$i_1 = 22\sqrt{2}\sin(\omega t - 53°)$$
$$i_2 = 22\sqrt{2}\sin(\omega t - 173°)$$
$$i_3 = 22\sqrt{2}\sin(\omega t + 67°)$$

7.2.2 三角形连接 (Delta-delta circuit)

负载做三角形连接时只能形成三相三线制电路,如图 7-6 所示。

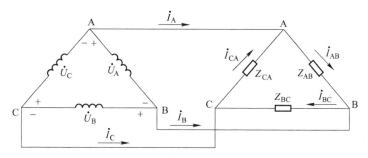

图 7-6 三角形连接

Fig. 7-6 Delta-delta circuit

当三相负载对称时,即各相负载完全相同,相电流和线电流也一定对称。

负载的相电流为:

$$\dot{I}_{AB} = \frac{\dot{U}_{AB}}{Z_A}$$

$$\dot{I}_{BC} = \frac{\dot{U}_{BC}}{Z_B} \qquad (7-16)$$

$$\dot{I}_{CA} = \frac{\dot{U}_{CA}}{Z_C}$$

线电流:

$$\dot{I}_A = \dot{I}_{AB} - \dot{I}_{CA} \qquad (7-17)$$

$$\dot{I}_B = \dot{I}_{BC} - \dot{I}_{AB} \qquad (7-18)$$

$$\dot{I}_C = \dot{I}_{CA} - \dot{I}_{BC} \tag{7-19}$$

根据 KCL：

$$\dot{I}_A + \dot{I}_B + \dot{I}_C = 0 \tag{7-20}$$

各相负载承受的电压均为对称的电源线电压。

在三相电源对称，三相负载也对称的情况下，线电流是对称的，相电流也是对称的。线电流的有效值为相电流有效值的$\sqrt{3}$倍，即

$$\dot{I}_{\Delta L} = \sqrt{3}\dot{I}_{\Delta P} \tag{7-21}$$

Note：

(1) In a delta-delta circuit, the line voltage is equal to the phase voltage.

(2) The relationship between the line current and phase current is $\dot{I}_{\Delta L} = \sqrt{3}\dot{I}_{\Delta P}$.

Example 7-2：

As it is shown in Fig. 7-6, find the currents in each phase of the loads and line currents. The load impedances are $Z = (30 + j10)\Omega$. The source voltages are $\dot{U}_{AB} = 1000\angle 30°$, $\dot{U}_{BC} = 1000\angle -90°$, and $\dot{U}_{CA} = 1000\angle 150°$.

Solution：

According to the Ohm's Law,

$$\dot{I}_{AB} = \frac{\dot{U}_{AB}}{Z} = \frac{1000\angle 30°}{30+j10} = \frac{1000\angle 30°}{31.62\angle 18.43°} = 31.63\angle 11.57°$$

$$\dot{I}_{BC} = \frac{\dot{U}_{BC}}{Z} = \frac{1000\angle -90°}{30+j10} = \frac{1000\angle -90°}{31.62\angle 18.43°} = 31.63\angle -108.43°$$

$$\dot{I}_{CA} = \frac{\dot{U}_{CA}}{Z} = \frac{1000\angle 150°}{30+j10} = \frac{1000\angle 150°}{31.62\angle 18.43°} = 31.63\angle 131.57°$$

According to the KCL,

$$\dot{I}_A = \dot{I}_{AB} - \dot{I}_{CA} = 31.63\angle 11.57° - 31.63\angle 131.57° = 54.78\angle -18.43°$$

$$\dot{I}_B = \dot{I}_{BC} - \dot{I}_{AB} = 31.63\angle -108.43° - 31.63\angle 11.57° = 54.78\angle -138.43°$$

$$\dot{I}_C = \dot{I}_{CA} - \dot{I}_{BC} = 31.63\angle 131.57° - 31.63\angle -108.43° = 54.78\angle 101.57°$$

7.3 三相功率（Three-phase power）

三相电路有功功率等于各相有功功率之和，三相电路无功功率等于各相无功功率之和。负载对称的话，每相有功功率都相等。因此，三相电路有功功率是各相有功功率的3倍。

$$P = 3P_p = 3U_p I_p \cos\varphi \tag{7-22}$$

式中 φ——某相相电压超前该相相电流的角度，即阻抗的阻抗角。

对称负载星形连接时：

对称负载三角形连接时：

$$U_1 = \sqrt{3}\,U_p, \qquad I_1 = I_p \tag{7-23}$$

$$U_1 = U_p, \qquad I_1 = \sqrt{3}\,I_p \tag{7-24}$$

则：

$$P = \sqrt{3}\,U_1 I_1 \cos\varphi \tag{7-25}$$

三相无功功率为：

$$Q = 3 U_p I_p \sin\varphi = \sqrt{3}\,U_1 I_1 \sin\varphi \tag{7-26}$$

三相视在功率为：

$$S = 3 U_p I_p = \sqrt{3}\,U_1 I_1 \tag{7-27}$$

Example 7-3：

For a wye connected load, the equivalent impedance is $Z = (29 + j21.8)\,\Omega$, and $U_1 = 380\text{V}$. Find the U_p, I_p, I_1 and P.

Solution：

$$U_p = \frac{U_1}{\sqrt{2}} = \frac{380}{\sqrt{2}} = 220\text{V}$$

$$I_p = \frac{U_p}{|Z|} = \frac{220}{\sqrt{29^2 + 21.8^2}} = 6.1\text{A}$$

In the wye connection：

$$I_1 = I_p = 6.1\text{A}$$

$$P = \sqrt{3}\,U_1 I_1 \cos\varphi = \sqrt{3} \times 380 \times 6.1 \times \frac{29}{\sqrt{29^2 + 21.8^2}} = 3200\text{W} = 3.2\text{kW}$$

7.4 练习题（Exercises）

7-1 In a wye-connected three-phase generator, find the value of phase voltage u_B, if the line voltage is $u_{BC} = 380\sqrt{2}\sin\omega t\,\text{V}$.

7-2 In a three-phase balanced wye-wye system, the sequence is ABCA, and the source is $\dot{U}_A = 220\angle 30°\,\text{V}$. The impedance of load of per-phase is $Z = (6+j8)\,\Omega$. If the line impedance per phase is $Z = (0.8 + j1.4)\,\Omega$, find the line currents and the load voltages.

7-3 In a three-phase balanced delta-delta system, the source has a positive ABC sequence. The line current is $\dot{I}_A = 15\angle 15°\,\text{A}$, and the load impedance is $Z = 10\angle 30°\,\Omega$. Find the total power absorbed by the load.

7-4 There are two balanced three-phase loads. One is wye connected, and the impedance of load of per-phase is $Z = (4 + j3)\,\Omega$. The other one is delta connected, and the impedance of load of per-phase is $Z = (10 + j10)\,\Omega$. The line voltage of the power source is 380V. Find the line currents of the wye connected load and delta connected load, respectively.

7-5 In a three-phase balanced delta load system, the line voltage of load is 380V, the phase current is 38A, and the power factor is 0.6. Find the real power, reactive power, and apparent power.

7-6　In a three-phase balanced delta load system, each phase has $X_R = 8\Omega$ and $X_L = 6\Omega$. The three-phase power source has $U_L = 380V$. Find the phase current and line current of the load.

7-7　A three-phase motor operating off a 400V system is developing 20kW at an efficiency of 0.87p.u. and a power factor of 0.82. Calculate the line current and the phase current if the windings are delta-connected.

7-8　The input power to a three-phase motor was measured by the two-wattmeter method. The readings were 5.2kW and −1.7kW, and the line voltage was 400V. Calculate (a) the total active power, (b) the power factor, and (c) the line current.

7-9　The load connected to a three-phase supply comprises three similar coils connected in star. The line currents are 25A, and the apparent and active power inputs are 20kV·A and 11kW, respectively. Find the line and phase voltages, reactive power input and the resistance and reactance of each coil. If the coils are connected in delta to the same three-phase supply, calculate the line currents and the active power taken.

7-10　In a three-phase four-wire system the line voltage is 380V, and non-inductive loads of 15kW, 10kW and 5kW are connected between the three line conductors and the neutral as in Fig. 7-7. Calculate: (a) the current in each line, and (b) the current in the neutral conductor.

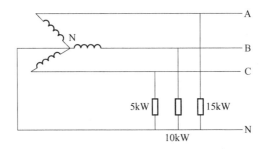

图 7-7　练习题 7-10

Fig. 7-7　Exercise 7-10

8 附录 (Appendix)

8.1 Appendix One: Expanded material

8.1.1 International system

The International System of Units (SI) is used as the legal system measurement in most countries in the world.

The SI is based on the measures of six physical quantities: mass, length, time, electric current, absolute temperature, and luminous intensity. All other units are derived units and are related to these base units by definition.

It is normal to use capital letters to represent constant quantities. In the case that the quantity varies, the lower case can be used. i.e. W indicates constant energy whereas w indicates a value of energy which is time varying.

The SI units adopt the abbreviation to make a convenient expression. The unit abbreviation is given in brackets.

e.g. Energy Symbol: W Unit: joule (J)

Table 8-1 Basic measures

Measure	Symbol	Unit	Unit symbol
mass	m	kilogram	kg
length	l/d	meter	m
area	A	square meter	m^2
volume	V	cubic meter	m^3
velocity	u	meter per second	m/s
acceleration	a	meter per second squared	m/s^2
angular velocity	v	radian per second	rad/s
force	F	newton	N
time	t	second	s
electric current	I	ampere	A
energy	W	joule	J

8.1.2 Electrical system

The role of an electrical system is to transmit energy from a source to the application. It is consists of four parts:

(1) Electrical energy production

(2) Electrical energy transmission

(3) Electrical energy application

(4) Electrical energy control

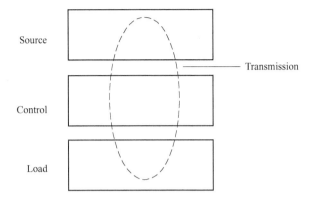

Fig. 8-1　Electrical system

Table 8-2　Electrical system

Component	Function	Example
source	provide the energy for the electrical system	battery or generator
load	absorb the electrical energy supplied by the source	lampor heater
transmission system	conduct the energy from the source to the load	insulated wire
control apparatus	control the status of energy flow	switch

8.1.3　Resistor

In an electrical circuit, the resistor owns the property named as resistance. For the linear resistor, its resistance is constant, and the resistor follows the Ohm's Law that the current through the resistor is proportional to the voltage across the resistor. For the non-linear resistor, the resistance varies with the change of current or the voltage. For example, the semi-conductor resistor is a typical type of non-linear resistor.

Concerning the aforementioned content, the resistor is linear. However, it is known that the resistance remains constant is under the condition of constant temperature. Therefore, it is assumed in this book that there is no effect of the temperature rise on the working resistor.

The resistor owns a power rating. It is the maximum power that can be dissipated without the temperature rise being such that damage occurs to the resistor. Thus, it should specify both the required resistance and power rating when we purchase a resistor. For instance, a resistor with 2W should be selected if a 1.1W resistor was required, as the rating must exceed the operational value.

In this book, it assumes that the resistor only has its according resistance given.

8.1.4　Series and parallel connected circuits

The comparisons of series and parallel-connected circuits are listed in Table 8-3.

Table 8-3 Series and parallel connected circuits

Property	Series-connected circuits	Parallel-connected circuits
circuit	$R_1, U_1, I_1 \quad R_2, U_2, I_2$	R_1, U_1, I_1 / R_2, U_2, I_2
current	$I = I_1 = I_2 = \cdots = I_n$	$I = I_1 + I_2 + \cdots + I_n$
voltage	$U = U_1 + U_2 + \cdots + U_n$	$U = U_1 = U_2 = \cdots = U_n$
resistance	$R = R_1 + R_2 + \cdots + R_n$	$\dfrac{1}{R} = \dfrac{1}{R_1} + \dfrac{1}{R_2} + \cdots + \dfrac{1}{R_n}$

Notes:

1. In a series connected circuit, the total resistance is always greater than each of the resistances.

2. In a parallel-connected network, the total resistance is always less than each of the resistances.

3. In a series arrangement, one load fails to work will interrupt the whole circuit. However, in the parallel-connected network, either load fails to work without affecting other branches.

8.1.5 Alternating system

Besides the circuits with direct current systems or transient systems, the alternating system is another significant system. It is defined by that the magnitudes of the voltage and of the current vary in a repetitive manner.

An alternating current flows first in one direction and then the direction varies. The variation cycle is repeated exactly for each direction.

The current waveforms relating with time are shown in Fig. 8-2, such as the sinusoidal wave, square wave, and triangular wave. The sinusoidal waveform is the most important AC waveform, as almost all electrical power supplies involve sinusoidal alternating current.

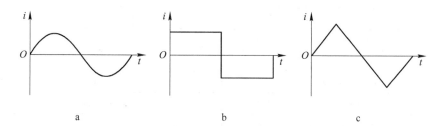

Fig. 8-2 Alternating current waveforms
a—sinusoidal wave; b—square wave; c—triangular wave

To describe the sinusoidal AC, there are a number of terms, such as the instantaneous value, maximum value, effective value, average value, period, frequency, angular frequency, and phase.

8.2　Appendix Two：Basic electric quantities and units

Table 8-4　Basic electric quantities and units

Quantity			Unit		
Symbol	Representation		Symbol	Representation	
U	voltage	电压	V	volt	伏特
u	instantaneous value		mV	millivolt	毫伏
U_m	maximum value		kV	kilovolt	千伏
I	current	电流	A	ampere	安培
i	instantaneous value		mA	milliampere	毫安
I_m	maximum value		μA	microampere	微安
E	electric field strength	电场强度	V/m	volt per metre	伏每米
q	charge	电荷	C	coulomb	库仑
f	frequency	频率	Hz	hertz	赫兹
T	period	周期	s	second	秒
t	time	时间			
W	energy	能量	J	joule	焦耳
			kJ	kilojoule	千焦
			W·h	watt hour	瓦特·小时
			eV	electronvolt	电伏特
P	power	功率	W	watt	瓦特
S	apparent power	视在功率	V·A	voltampere	伏·安
F	force	力	N	newton	牛顿
R	resistor	电阻	Ω	ohm	欧姆
Z	impedance	阻抗			
X	reactance	电抗			
X_C	capacitive reactance	容抗			
X_L	inductive reactance	感抗			
Y	admittance	导纳	S	siemens	西门子
G	conductance	电导			
ω	angular velocity	角速度	rad/s	radian per second	弧度/秒
s	conductivity	电导率	S/m	siemens per meter	西门子/米
L	inductance	电感	H	henry	亨利
C	capacitance	电容	F	farad	法拉
Q	reactive power	无功功率	var	var	乏尔
T	torque	力矩	N·m	newton·metre	牛顿·米
τ	time constant	时间常数			

8.3 Appendix Three: Multiples/sub-multiples abbreviations

Table 8-5 Multiples/sub-multiples abbreviations

Abbreviation	Name	Relationship
Y	yotta	10^{24}
Z	zetta	10^{21}
E	exa	10^{18}
P	peta	10^{15}
T	tera	10^{12}
G	giga	10^{9}
M	mega or meg	10^{6}
k	kilo	10^{3}
h	hecto	10^{2}
da	deca	10
d	deci	10^{-1}
c	centi	10^{-2}
m	milli	10^{-3}
μ	micro	10^{-6}
n	nano	10^{-9}
p	pico	10^{-12}
f	femto	10^{-15}
a	atto	10^{-18}
z	zepto	10^{-21}
y	yocto	10^{-24}

8.4 Appendix Four: Mathematic representation

Table 8-6 Mathematic representation

Term	Symbol
+	plus
−	minus
×	multiple
÷	divide
=	equal to
≃	approximately equal to
∝	proportional to
∞	infinity
Σ	sum of
Δ/d	increment or finite difference operator
>	greater than
<	less than
e	base of natural logarithms
$\lg x$	common logarithm of x
$\ln x$	natural logarithm of x

8.5 Appendix Five: Greek letters

Table 8-7 Greek letters

Letter	Capital	Lower case
alpha	A	α
beta	B	β
delta	Δ	γ
epsilon	E	ε
eta	H	η
theta	Θ	θ
lambda	Λ	λ
mu	M	μ
pi	Π	π
rho	P	ρ
sigma	Σ	σ
phi	Φ	φ
psi	Ψ	ψ
omega	Ω	ω

8.6　Appendix Six：Vocabulary

A

abbreviation 缩写语
AC 交流电
alternating current 交流电
amplify 放大、增强
amplitude 振幅
analysis 分析
angular frequency 角频率
anode 阳极
apparent power 视在功率
average power 平均功率

B

balanced 对称的
balanced three-phase system 三相对称系统
bracket 括号
branch 支路
bronze 青铜
bus 母线
button 按钮

C

capacitance 电容
capacitor 电容器
cathode 阴极
charge 电荷
circuit 电路
conduct 导电
conductivity 导电性
constant amplitude 等幅振荡
coil 线圈
contactor 接触器
controlled source 受控电源
copper 铜
core 芯
corresponding 相应的
current 电流

D

delta connection 三角形连接
delta-connected load 三角形连接负载
dielectric 电介质、绝缘体
dielectric constant 介电常数
digital circuit 数字电路
direct current 直流电
direction 方向
discharge 放电
display 显示器
divide 除

E

electric 电的
electric power 电功率
electric potential 电位
electron 电子
electromotive force 电动势
element 元件
equation 方程

F

force 力
frequency 频率

G

glass 玻璃

H

Henry 亨利
hot rail 火线
hybrid 混合的

I

impedance 阻抗
in phase 同相
indicator 指示器

inductor 电感器
inductance 电感
instantaneous value 瞬时值
integrated 集成的
iron 铁
iron core 铁芯

J
junction 节

K
Kirchhoff's Current Law 基尔霍夫电流定律
Kirchhoff's Voltage Law 基尔霍夫电压定律

L
ladder diagram 梯形图
lag 滞后
lead 超前
linear resistance 线性电阻
limit switch 限位开关
loop 回路

M
magnetic circuit 磁路
magnetic field strength 磁场强度
magnetomotive force 磁动势
magnitude 大小
mesh 网孔
minus 减
multiply 乘积

N
neutral 中性的
neutral point 中性点
no-load 空载
nonlinear resistance 非线性电阻
norton theorem 诺顿定理
nodal 节点的
node 节点

network 网络

O
Ohm's law 欧姆定律
on-load 有载
open circuit 开路
opposite in phase 反相
overload ratio 过载系数

P
paper 纸
parallel circuit 并联电路
parameter 参数
phase 相
phase angle 相位角
phase sequence 相序
phase voltage 相电压
phasor diagram 相量图
plus 加
positive phase sequence 正相序
power 功率
power factor 功率因数
power triangle 功率三角形
power source 电源
problem 问题

Q
quality 质量
quality factor 品质因数

R
radian 弧度
ratio 比率
reactance 电抗
reactive power 无功功率
real power 有功功率
resistance 电阻
resistor 电阻器
resonance 谐振

reverse 反转
root mean square, RMS 均方根（有效值）
rubber 橡胶

S
semiconductor 半导体
sequence 顺序
series circuit 串联电路
short circuit 短路
sinusoidal 正弦的
speed control 调速
steady state 稳态
superposition principle 叠加原理
switch 开关

T
terminal voltage 端电压
three elements method 三要素法
Thévenin's theorem 戴维南定理
time constant 时间常数

torque 转矩
transient 暂时的
turns 匝数

U
U 电压符号

V
volt 伏特
voltage 电压

W
winding 绕组
wood 木
wye connection 星形连接
wye-connected load 星形连接负载

Z
Z 阻抗符号

参考文献（References）

[1] 忻尚芝. 电工与电子电路 [M]. 上海：上海科学技术出版社，2016.
[2] 汤春明，马惠珠，张忠民. 电工技术：电工学（上册）[M]. 北京：清华大学出版社，2011.
[3] 王逸隆. 电工基础 [M]. 北京：化学工业出版社，2013.
[4] 江路明. 电路分析与应用 [M]. 北京：高等教育出版社，2015.
[5] 徐存善. 机电专业英语 [M]. 北京：机械工业出版社，2016.
[6] Europa Lehrmittel 出版社组. 电气工程学 [M]. 北京：机械工业出版社，2013.
[7] John Hiley, Keith Brown, Ian Mckenzie Smith. Electrical and Electronic Technology [M]. Harlow：Pearson Education Limited，2016.